Introduction to Environmental Impact Assessment

A Guide to Principles and Practice

Bram F. Noble

OXFORD
UNIVERSITY PRESS

OXFORD
UNIVERSITY PRESS

70 Wynford Drive, Don Mills, Ontario M3C 1J9
www.oup.com/ca

Oxford University Press is a department of the University of Oxford.
It furthers the University's objective of excellence in research, scholarship,
and education by publishing worldwide in

Oxford New York

Auckland   Cape Town   Dar es Salaam   Hong Kong   Karachi
Kuala Lumpur   Madrid   Melbourne   Mexico City   Nairobi
New Delhi   Shanghai   Taipei   Toronto

With offices in

Argentina   Austria   Brazil   Chile   Czech Republic   France   Greece
Guatemala   Hungary   Italy   Japan   Poland   Portugal   Singapore
South Korea   Switzerland   Thailand   Turkey   Ukraine   Vietnam

Oxford is a trade mark of Oxford University Press
in the UK and in certain other countries

Published in Canada
by Oxford University Press

Copyright © Oxford University Press Canada 2006

The moral rights of the author have been asserted

Database right Oxford University Press (maker)

First published 2006

Library and Archives Canada Cataloguing in Publication

Noble, Bram F., 1975-
Introduction to environment impact assessment : guide to
principles and practice / Bram F. Noble

Includes bibliographical references.
ISBN-13: 978-0-19-542090-6.-   ISBN-10: 0-19-542090-X

1. Environmental impact analysis—Textbooks. 2. Environmental
impact analysis—Canada—Textbooks. I. Title.

TD194.6.N62 2005   333.71'4'0971   C2005-904751-8

Cover Design: Brett Miller

1 2 3 4 – 09 08 07 06
This book is printed on environmentally responsible Rolland Enviro 100.
Printed in Canada

To Deana, Noelle, and Elias

Contents

List of Boxes

List of Tables

List of Figures

Preface: Using This Guide

This book is directed to beginning students of environmental impact assessment (EIA). It is written primarily for introductory courses on the practice of EIA, in both academic and professional environments. Current practitioners and administrators may find it particularly useful as a reference book for key EIA principles, design, and concepts.

It is not my intent here to explore in detail Canadian or international experiences with EIA application—a text that is sufficiently comprehensive for EIA procedural training should not at the same time attempt to examine in detail the regulatory requirements, experiences, and countless case studies related to impact assessment. This would become cumbersome, and my focus here is on EIA process. That said, however, the book does include numerous case studies, primarily from Canadian experience but also drawn from American, European, and Australian contexts.

The main focus of this book is on the fundamental principles and 'good' practices of EIA that are, arguably, common across EIA systems. The intent is that the course instructor will use this book as a framework and integrate case examples particular to the specific EIA system of concern. End-of-unit questions and exercises are designed to facilitate this approach. In recognition that EIA is a subject of interest to many disciplines and persons of varied backgrounds, this book is interdisciplinary in perspective and attempts to balance discussion on physical and human environments.

The book is organized in four parts, plus appendices and a glossary. In each part a number of principles and procedures are introduced, including case study examples and select methods and techniques. Part I introduces the basic principles of EIA, including the aims and objectives of EIA and an overview of the generic EIA process. Part II examines the nature of environmental impacts and common methods and techniques used to support EIA practice. Part III focuses on EIA principles and procedures, from screening, scoping, impact prediction, and evaluation to post-decision analysis. Part IV introduces two key areas of EIA that are under rapid transition—cumulative effects assessment and strategic environmental assessment—and demonstrates linkages between the two and linkages with project-based EIA.

Each of the 12 chapters concludes with a list of key terms, which are defined in the Glossary at the end of the book, and study questions and exercises. Many of the questions and exercises, as well as the basic principles and procedures of EIA, are consolidated in the two appendices, which are practical exercises in project and strategic assessment. These exercises are hypothetical, but are designed to be tailored to the particular setting and jurisdiction within which this book is being used. The objective of the exercises is to provide course participants with an opportunity to apply the concepts and procedures demonstrated throughout the book. The exercises may be useful as a term project in academic settings, or as a day-long workshop exercise in professional training environments.

Bram F. Noble
Department of Geography
University of Saskatchewan

Acknowledgements

I am grateful to the staff at Oxford University Press for their assistance throughout the manuscript writing and editing process. I am particularly grateful to Richard Dixon for encouraging me to undertake this project in the first place. I would also like to acknowledge my winter 2004 Geography 386 class at the University of Saskatchewan, who used the first draft of this manuscript as part of the course materials and provided valuable feedback and suggestions for improvement. Two students at the University of Saskatchewan, Kimberely Janzen and Jackie Bronson, also provided assistance with manuscript research, case study selection, and manuscript styling.

ENVIRONMENTAL IMPACT
ASSESSMENT PRINCIPLES

CHAPTER 1

Aims and Objectives of Environmental Impact Assessment

INTRODUCTION

Environmental, economic, and social changes are inherent to human development. While the goal of development is typically positive change, it can also create potentially negative impacts on the natural and human environment. The challenge facing governments and industries is to find ways to support development initiatives while enhancing health and well-being without adversely affecting the environment. Managing the impacts of human activities on the environment, as well as creating positive environmental outcomes from such activities, is essential if development is to be recognized as sustainable. The interest in environmental issues has been unmatched in recent years, as illustrated by the introduction of new state environmental legislation. With growing populations, technological advancement, and increasing demand for natural and human resources, the need for common tools and mechanisms to manage effectively the impacts of development is of ever-increasing importance.

DEFINING ENVIRONMENTAL IMPACT ASSESSMENT

According to the World Bank, **environmental impact assessment** (EIA) is the most widely practised environmental management tool in the world. There is no single universally accepted definition of environmental impact assessment (Box 1.1), and EIA is often defined as a 'tool', a 'methodology', and a 'regulatory requirement'. Arguably, EIA is all of these; but perhaps most importantly it is a 'process' designed to aid decision-making through which concerns about the potential environmental consequences of proposed actions, public or private, are incorporated into decisions regarding those actions. The International Association for Impact Assessment (IAIA) and the UK Institute of Environmental Assessment (IEA), for example, define EIA as:

> The process of identifying, predicting, evaluating and mitigating the biophysical, social, and other relevant effects of development proposals prior to major decisions being taken and commitments made. (IAIA and IEA, 1999)

Box 1.1   Common Definitions of Environmental Impact Assessment

EIA evaluates all relevant environmental and resulting social effects which would result from a project (Battelle, 1978).

EIA is an activity designed to identify and predict the impact on human health and well-being of legislative proposals, policies, programs and operational procedures, and to interpret and communicate information about the impacts (Munn, 1979).

EIA is the study of the full range of consequences, immediate and long range, intended and unanticipated, of the introduction of a new technology, project or program (Rossini and Porter, 1983).

EIA is a tool for identifying and predicting the impacts of projects and investigating and proposing means for their management (CEARC, 1988).

EIA is a planning tool whose main purpose is to give the environment its due place in the decision-making process by clearly evaluating the environmental consequences of a proposed activity before action is taken (Gilpin, 1995).

More specifically, EIA is:

- a comprehensive and systematic process designed to identify, analyze, and evaluate the environmental effects of proposed projects;
- a process that allows for the effective integration of environmental considerations and public concerns into decision-making processes; and
- a powerful tool to help decision-makers achieve the goal of sustainable development.

EIA is an organized means to gather information used to identify and understand the effects of proposed projects on the **biophysical environment** (air, water, land, plants, and animals) as well as on the **human environment** (culture, health, community sustainability, employment, financial benefits) for those people potentially affected. In simplest terms, EIA is an integral component of sound decision-making, serving both an information-gathering and analytical component, used to inform decision-makers concerning the management of proposed developments.

EIA Objectives and Principles

The primary purpose of EIA is to facilitate the consideration of environment in planning and decision-making and, ultimately, to make it possible to arrive at decisions and subsequent actions that are more environmentally sustainable. In this regard, EIA should not be seen merely as a mechanism for preventing development that might generate potentially negative environmental impacts. If this were the case, then few developments would actually take place! Rather, EIA is intended to:

- systematically identify and predict the impacts of proposed development;

- find ways to avoid or minimize significant negative biophysical and socio-economic impacts;
- identify, enhance, and create potentially positive impacts; and
- ensure that development decisions are made in the full knowledge of their environmental consequences.

The objectives of EIA then can be separated into 'output objectives' and 'outcome objectives'. Output objectives are the immediate, short-term objectives of the EIA process and include:

- ensuring that environmental factors are explicitly addressed in decision-making processes concerning proposed developments;
- improving the design of the proposed development;
- anticipating, avoiding, minimizing, and offsetting adverse effects relevant to development proposals on the human and biophysical environment; and
- facilitating informed decision-making.

Outcome objectives are the longer-term objectives that are the products of consistent and rigorous EIA application. Such objectives are realized from a broader environmental and societal perspective, and include:

- protecting the productivity and capacity of human and natural systems and ecological functions;
- protecting human health and safety;
- facilitating participatory approaches to development; and
- promoting development that is sustainable.

For example, an immediate output of EIA could be the abandonment of environmentally unacceptable actions and mitigation, to the point of acceptability, of the environmental effects of proposals that are approved. A longer-term outcome of EIA would thus be an overall positive contribution to the environment and society.

In practice, the underlying objectives of EIA are based on three *core values* (Sadler, 1996): integrity, sustainability, and utility. Integrity implies that EIA should conform to standards and general principles of 'good practice'. Sustainability suggests that EIA should promote environmentally sound and socially acceptable development. Utility is based on the premise that EIA should provide balanced and credible information for planners and decision-makers to make informed decisions. Together, the desired outcome, according to Sadler, is that an EIA results in information that is accurate and appropriate to the nature of development, identification of likely impacts, desired outcomes, and strategies for environmentally sound management practices. These core values have been expanded upon by the IAIA and the IEA (1999) in the form of a comprehensive set of *basic principles* for EIA based on the notion of 'best practice' (Figure 1.1). These principles apply to all stages of EIA and are meant to be interpreted as a single package, and they are both interdependent and necessary to ensure that EIA fulfills its intended objectives. At the same time, what is 'best practice' in one situa-

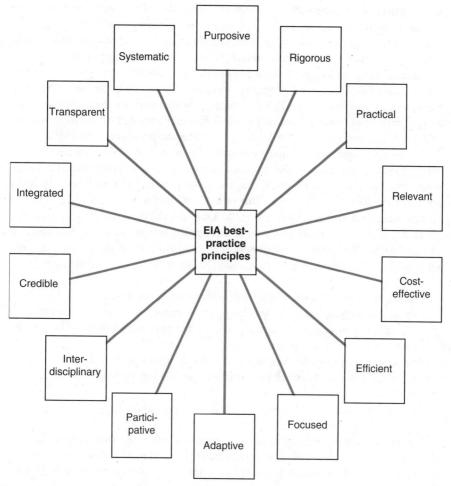

Figure 1.1 Basic Principles of 'Best-Practice' EIA

Source: Based on IAIA and IEA (1999).

tion may not necessarily be so in another. 'Best practice' simply means adhering to a number of general principles, or the best way of doing things, as determined by the specific social, political, cultural, and environmental context within which the EIA is taking place.

THE ORIGINS AND DEVELOPMENT OF EIA

EIA Origins

Throughout North America and Western Europe, the 1960s were characterized by a sudden growth in awareness of the relationship between an expanding industrial economy and local environmental change. While many characterize the 1960s as an era of environmental 'idealism', exacerbated by a number of environmental challenges and sparked by such works as Rachel Carson's *Silent Spring* (1962), the decade did lead to increasing environmental awareness and public demand and pressure on central governments that environmental factors be explicitly considered in development decision-making. As a result of such pressures, the 1960s and early 1970s witnessed the passage of legislation concerning resource protection, hazardous waste management, and control of water and air pollution. However, perhaps the most significant piece of legislation was the **National Environmental Policy Act** (NEPA) in the United States, which came into effect in January 1970 after its introduction in 1969.

The term 'environmental impact assessment' is actually derived from NEPA, which for the first time required by law that those proposing to undertake certain development projects had to demonstrate that the projects would not adversely affect the environment. To do so, the proponents of a project had to include in their proposal an **environmental impact statement** (EIS) describing the proposed development, the affected environment, likely impacts, and proposed actions to manage or monitor those impacts. The specific objectives of the initial EIA process under NEPA are outlined by the US Council on Environmental Quality (CEQ, 1973), namely that:

- each generation is to serve as a trustee for future generations;
- a safe, productive, and socially pleasing environment is to be maintained;
- a wide range of safe and acceptable uses of the environment should be permitted;
- heritage and options for future generations should be protected;
- a balance between resource potential and population growth must be encouraged;
- renewable and recyclable resource use is to be encouraged.

During the first 10 years of NEPA, approximately 1,000 EISs were prepared annually. Currently, it is estimated that approximately 30,000 to 50,000 EISs are prepared annually in the US, and some form of EIA system exists in more than 30 individual states.

Prior to the 1970s development projects were still being assessed, but assessment was limited to technical feasibility studies and, in particular, **cost-benefit analysis**. Cost-benefit analysis (CBA) is an approach to project assessment that expresses impacts based on monetary terms. Large-scale development projects such as the 114-metre-high Aswan High Dam in Egypt, for example, constructed from 1960 to 1970 and financed by the US and Great Britain, were assessed using CBA techniques. The US Corps of Engineers similarly used CBA for many years to assess and justify large-scale water resource development projects and dam construction in the United States, including the Glen Canyon Dam, Arizona (completed in 1964), and the Oroville Dam in California (completed in 1968). While still commonly used today, obvious drawbacks to CBA include the inability to allocate meaningful dollar values to environmental and human intangibles and the limited scope of fiscal impacts traditionally addressed by CBA. As an 'add-on process' to CBA, EIA was initially intended to incorporate all those potential impacts that were excluded from traditional CBA.

EIA Development

The US NEPA is generally recognized as the pioneer of contemporary environmental impact assessment. Since its beginnings EIA has gone through a number of evolutionary phases in the United States and Canada, a pattern repeated to varying degrees throughout the world (see Wood, 1995; Sadler, 1996).

Initial development. During the early 1970s, immediately following the implementation of NEPA, EIA was characterized by casual and disjointed observations of the biophysical environment, particularly within the local project area. Wider project impacts and potential impact interactions among physical and human environmental components were largely ignored. During this phase EIA was primarily criticized as a tool used to justify project decisions already made when the process was initiated. The ethos that predominated this era was one of 'develop now, minimize associated costs and, if forced to, clean up later' (Barrow, 1997). In many cases, as illustrated in Chapter 2, lands were leased and project construction was well underway before EIA started. The end result was that by the time EIA commenced many of the opportunities to integrate environmental concerns into project planning and to identify more environmentally sustainable courses of action were foreclosed.

Broadening scope and techniques. By the mid-1970s and up to the early 1980s, EIA efforts became much more highly organized and technically oriented, reflecting the interests of professionals in the natural sciences. Significant attention was devoted to collecting large environmental inventories, i.e., comprehensive descriptions of the biophysical environment in local project areas. The result was impact statements that were thousands of pages in length, often consisting of little more than a compilation of biophysical environmental facts. It is perhaps not surprising, then, that during this time project **scoping** was first introduced in an attempt to identify the important issues and data requirements for project assessment. Greater attention turned towards managing adverse project impacts and the risks associated with particular actions, as opposed to creating large environmental inventories. Public review of project proposals and impact assessment processes also emerged as common practice in EIA, and several new innovative impact prediction and assessment techniques were introduced. Perhaps more importantly, EIA advanced beyond the biophysical environment of the local project in recognition that broader regional and social impacts were also important when evaluating the impacts of proposed project actions on the environment.

Institutional support and integration. The early 1980s to the mid-1990s witnessed rapid growth in EIA, attributed to, among other things, a number of international events such as the 1987 World Commission on Environment and Development and the 1992 and 1997 Earth Summits, all of which fostered greater international awareness of EIA. During this period, EIA practice turned increasing attention towards physical and social interrelationships associated with project development. **Environment** within the context of EIA was defined to be inclusive of not only the biophysical environment, but also components of the social and economic environments (e.g., labour markets, demography, housing structure, education, health, values, lifestyles) at multiple spatial and temporal scales (Figure 1.2). In 1987, for example, the World Health Organization introduced 'environmental health impact assessment' in recognition that for certain projects, such as nuclear power plants and large industrial complexes,

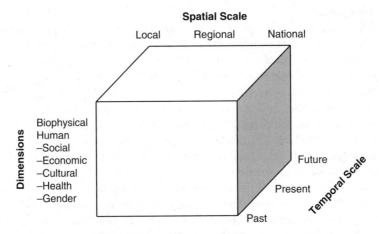

Figure 1.2   Dimensions of 'Environment' in EIA

it was necessary to address broader environmental, physical, social, and psychological health impacts in project decision-making.

EIA thus emerged in the 1990s as a system-oriented or multi-dimensional approach, and involved applications of both qualitative and quantitative models in impact prediction and analysis. Emphasis increasingly turned to the assessment of 'total environmental impacts' and the inclusion of environmental considerations early in the project planning cycle before irreversible decisions were made. This gave way to greater attention to the links between EIA and impact management, social effects assessment, health impact assessment, and adaptive environmental management, and led to increasing awareness of the need for better monitoring of environmental effects after project implementation. Originally conceived in reaction to growing environmental awareness and to the voice of environmental lobby groups during the late 1960s, EIA was emerging as an increasingly important environmental management tool.

Sustainability initiatives. There is now a growing recognition that EIA should serve as an integrated planning tool for decision-making, characterized by integrating cumulative and global environmental effects, empowering the public, recognizing uncertainties, favouring a precautionary and adaptive approach, and making a positive contribution towards sustainability (Gibson, 2002). One might say that the breadth of EIA has spread more than its geographical extent. To that end, Richard Morgan, president of the International Association for Impact Assessment (IAIA), commented in a recent newsletter of the IAIA on the disillusionment with measures to promote sustainable development and skepticism that EIA is in fact contributing to better decisions (Morgan, 2005). Morgan suggests that there may be too many different things expected from EIA, and that there perhaps are too many different ideas as to what EIA can accomplish. EIA is not a 'magic bullet'—while it complements broader environmental planning, management, and decision-making, it does not replace them. Such disillusionment and skepticism may, in part, be due to the fact that the fundamental principals of EIA (Figure 1.1) are not being respected. While examples exist of Canadian EIA that are set explicitly within the context of sustainability, such as the Voisey's Bay nickel mine-mill project (Box 1.2), for the most part EIA is still viewed primarily as a tool for preventing or minimizing environmental problems. In many contexts, EIA has not yet fully advanced beyond this phase and in many countries EIA is still not done prior to the design and approval of projects or plans.

Advancing the sustainability initiative will require increasing the application of EIA principles beyond the project level to address environmental issues at the strategic levels of policy, planning, and program decision-making. This can be accomplished through **strategic environmental assessment** (SEA), the application of environmental assessment principles to policies, plans, and programs. SEA is based on the notion that the benefits of sustainable development trickle down from policy decisions to plans, programs, and eventually to individual projects. While SEA is increasingly recognized as a critical tool for advancing the sustainability agenda, considerable

---

Box 1.2 Voisey's Bay Mine-Mill Project

In 1993 a rich nickel-copper-cobalt deposit was discovered at Voisey's Bay in northern coastal Labrador. The deposit has surface dimensions of approximately 800 metres by 350 metres, extends to depths of about 125 metres, and will be mined using open-pit methods. The deposit contains estimated proven and probable mineral reserves of 30 million tonnes, grading 2.85 per cent nickel, 1.68 per cent copper, and 0.14 per cent cobalt. There are an additional estimated 54 million tonnes of mineral wealth at Voisey's Bay, grading 1.53 per cent nickel, 0.70 per cent copper, and 0.09 per cent cobalt, as well as 16 million tonnes of inferred mineral resource, grading 1.6 per cent nickel, 0.8 per cent copper, and 0.1 per cent cobalt.

The project proponent, Voisey's Bay Nickel Company Limited (VBNC), a subsidiary of Inco Limited, submitted in 1997 a proposal for the development of a mine-mill complex and related infrastructure to produce mineral concentrates at Voisey's Bay. In the absence of Aboriginal land claims agreements, the Voisey's Bay project was subject to review as set out under Canadian federal and Newfoundland provincial environmental assessment processes and pursuant to a memorandum of understanding between the provincial and federal governments, the Labrador Inuit Association, and the Innu Nation.

The public review panel for the Voisey's Bay project issued *Guidelines for Review* in which the proponent was required to discuss explicitly the extent to which the project would 'make a positive overall contribution towards the attainment of ecological and community sustainability, both at the local and regional levels' (Voisey's Bay Panel, 1997). This was the first major resource development project in Canada for which the impact statement guidelines for the project proponent explicitly identified the sustainability criterion, noting that EIA should go beyond seeking to minimize damage to require that an undertaking maximize long-term, durable net gains to the community and the region. Construction of the mine-mill project commenced in 2002. See <www.vbnc.com/> for additional project information.

process development is still required. The principles and practice of SEA are the focus of discussion in Chapter 12.

INTERNATIONAL STATUS OF EIA

Since NEPA, EIA has diffused throughout the world, and is currently applied in over 100 countries (Box 1.3). While NEPA provided the initial basis for EIA, every nation's EIA system is quite distinct and the enabling legislation, policy directives, and guidelines for EIA vary considerably from one nation to the next. It is not possible to review the status of EIA in all nations, so only a few national EIA systems are highlighted here. A more in-depth discussion of national EIA systems can be found in Glasson et al. (1999). It is important to note that while EIA is expanding internationally, it is still relatively new as many nations have less than 10 years of EIA experience.

Box 1.3 Selected National EIA Systems

Country	Date First Implemented*	Web Site to Current System
Armenia	1995	www.mnpiac.am
Australia	1974	www.ea.gov.au
Bulgaria	1991	www.moew.government.bg
Brazil	1981	www.mma.gov.br/
Canada	1973	www.ceaa.gc.ca
Chile	1993	www.conama.cl/portal/1255/channel.html
China	1989	www.zhb.gov.cn/english/
Colombia	1974	www.minambiente.gov.co/
France	1976	www.environnement.gouv.fr
Germany	1985	www.bmu.de
Guyana	1996	www.epaguyana.org/
Israel	1982	www.sviva.gov.il
Japan	1984	www.env.go.jp/en/index.html
Madagascar	1995	www.madonline.com/one
New Zealand	1974	www.mfe.govt.nz/
Pakistan	1983	www.punjab.gov.pk/epa/index.htm
South Africa	1984	www.environment.gov.za/
Thailand	1975	www.thaigov.go.th/index-eng.htm
United States	1969	www.epa.gov/

*Many nations have undergone revisions in their EIA system since first introduced, including moving from guidelines or policy directives to formal legislated systems. Canada, for example, introduced EIA in 1973, but it was not until 1995 that EIA was formally legislated. See also the following Web site for a directory of environmental agencies of the world: <www.inece.org/links_pages/onlineresourcesEnvironmentalagencies.html>.

Canada

Canada was first to follow the US NEPA beginnings, formally implementing an EIA system in 1973 as a guidelines order through the federal Environmental Assessment and Review Process. It was not until 1995, however, that EIA became entrenched in Canadian law, and subsequent amendments have since taken place, including a 2003 amendment to the Canadian Environmental Assessment Act. Thousands of EIAs are completed each year in Canada, ranging from local initiatives to large-scale resource development projects. Currently, all Canadian provinces have their own EIA systems with additional EIA systems for Aboriginal land claim areas. The Canadian EIA system is discussed in greater detail in Chapter 2.

Australia

Australia was next to formally adopt a system of EIA in 1974 through its Environmental Protection Act. The Act was formally implemented in 1975 and amended in 1995. Similar to Canada, most states in Australia have opted for their own form of EIA legislation where projects that are of national significance are assessed under a joint state and federal system. The largest portion of national-level EIAs completed in Australia concern applications for mineral resource development, followed by transport and military developments (Glasson et al., 1999).

European Union

EIA systems were first introduced in France in 1976, which accounts for by far the most EIAs in Europe, averaging between 5,000 and 6,000 per year (ibid.). It was not until 1985, however, that EIA was formally adopted in Europe through European Union Directive 85/337/EEC, amended in 1997, which set a legal basis for individual member states' EIA regulations (Morris and Therivel, 2001). In 1991 the Espoo Convention, signed by the European Commission and several UN Economic Commission for Europe member states, provided a framework for EIA to address potential transboundary impacts. By the mid-1990s more than 2,500 EISs had been produced in the UK, mostly from England for waste disposal projects, followed by industrial and urban developments, with an average of approximately 300 EISs produced each year (Glasson et al., 1999).

Japan

EIA was discussed in Japan as early as the mid-1970s and a bill was proposed to parliament in 1976, but it was not until 1984 that EIA was formally adopted through non-mandatory guidelines. Glasson et al. (1999) report that more than half of Japan's local governing authorities have adopted some form of EIA, all of which are more stringently regulated than the national guidelines. A formal legislated EIA system was introduced in Japan in 1997. There are currently about 70 EIAs prepared in Japan each year.

China

In China, EIA is similarly implemented through guidelines rather than legislation. A 1979 Environmental Protection Law provided the basis for EIA, which was subsequently amended in 1989. EIA in China is currently governed by a National Environmental Protection Agency and by more than 2,000 regional and municipal Environmental Protection Bureaus, which receive thousands of EIAs annually.

Developing Nations and Development Agencies

There are now more than 70 developing or transitional nations with EIA systems in place. Perhaps the region where the greatest political transition is currently taking place is Africa, which has EIA systems operating in about a dozen countries, most of which have been established within the past decade. Many of these systems operate on an ad hoc basis in response to the requirements of international donor agencies, such as the World Bank. The World Bank first introduced EIA requirements in 1989 for evaluating projects it was financing; the Asian Development Bank followed in 1993 with similar requirements (Harrop and Nixon, 1999). Similarly, the Canadian International Development Agency (CIDA) has EIA requirements for international investment and development projects, and recently adopted a system of environmental assessment to address the potential environmental impacts associated with policy and program decisions.

EIA PROCESS

From an applied perspective, EIA can be thought of as a *process* that systematically examines the potential environmental implications of development actions prior to project approval. In short, EIA is simply responsible and proactive planning. The structure of an EIA process, however, is dictated by the specific issues it attempts to address and by the regulatory and legislative requirements within which it operates. While not all EIA systems contain the same elements or specific design procedure, the broad EIA process emanating from the original NEPA and subsequently diffused throughout the world can be thought of as a series of systematic steps (Box 1.4). Although presented here as a linear framework, in practice EIA is an iterative process in which discussions with stakeholders, public review, scoping processes, and post-project evaluations continue to refine impact predictions and management actions. The various stages of the EIA process are explored in greater detail throughout this book.

Several *operating principles* of EIA define how the EIA process should unfold. It should be applied:

- as early as possible in the planning and decision-making stages;
- to all proposals that may generate significant adverse effects or for which there is significant public concern;
- to all biophysical and human factors potentially affected by development, including health, gender, and culture, and cumulative effects;
- consistently with existing policies, plans, programs, and the principles of sustainable development;

Box 1.4   Generic EIA Process

Project description	Description of the proposed action, including its alternatives, and details sufficient for an assessment.
Screening	Determination of whether the action is subject to an EIA under the regulations or guidelines present, and if so what type or level of assessment is required.
Scoping	Delineation of the key issues and the boundaries to be considered in the assessment, including the baseline conditions and scoping of alternatives.
Impact prediction and evaluation	Prediction of environmental impacts and determination of impact significance.
Impact management	Identification of impact management and mitigation strategies, and development of environmental management or protection plans.
Review and decision	Technical and public review of EIS and related documents, and subsequent recommendation as to whether the proposed action should proceed and under what conditions.
Implementation and follow-up	Implementation of project and associated management measures; continuous data collection to monitor compliance with conditions and regulations; monitoring the effectiveness of impact management measures and the accuracy of impact predictions.

Public consultation

- in a manner that allows involvement of affected and interested parties in the decision-making process;
- in accordance with local, regional, national, or international standards and regulatory requirements (IAIA and IEA, 1999).

How the process actually unfolds in practice and the quality of its application, however, vary considerably from one nation to the next (Box 1.5). Easily recognizable from Box 1.5 is the poor performance of impact monitoring across most national EIA systems. This raises an important question: How does one determine the effectiveness of EIA and the measures put in place to manage impacts in the absence of impact monitoring programs? This issue is returned to in Chapter 10.

BENEFITS AND CHALLENGES TO EIA

In addition to managing the impacts of development on the environment, there are a number of direct benefits to EIA, including:

Box 1.5 Performance of Selected National EIA Systems

Evaluative Component	Australia	Canada	UK	US	New Zealand	Netherlands
Screening	☒	☑	☑	☑	☑	☑
Scoping	☑	☑	☒	☑	☐	☑
Content of EIS	☑	☑	☐	☑	☒	☑
Mitigation	☑	☑	☑	☑	☑	☑
Impact monitoring	☒	☐	☒	☒	☒	☐
Public consultation	☐	☐	☐	☑	☐	☑

☑ Good (criterion met)

☐ Average performance (criterion partially met)

☒ Insufficient (criterion not met)

Source: Based on Petts (1999).

- improvements to project planning and design;
- cost savings for proponents by early identification of potentially unforeseen impacts;
- reducing the role of legality by ensuring early compliance with environmental standards;
- increasing public acceptability through participation and demonstrated environmental and socio-economic responsibility.

That said, a number of challenges to EIA still must be addressed if EIA is to fulfill its role as a process towards sustainability. Such challenges are reflected in Trudgill's (1990) 'barriers to a better environment', which identify a number of common barriers that stand in the way of environmental problem-solving. While Trudgill's work was set primarily within the context of rain forest depletion and acid rain, the framework does capture those barriers to effective EIA, namely:

1. Agreement barrier: agreement among various interests and practitioners as to what EIA can and should accomplish.
2. Knowledge barrier: knowledge about the EIA process, its potential value, its role within planning and project decision-making, and knowledge about the *actual* impacts of projects on the environment post-project implementation.
3. Technology barrier: availability and appropriate use of methods and techniques to improve the accuracy and precision of project impact prediction and the effectiveness of impact mitigation measures.

4. Economic barrier: the financial cost of doing EIA, including the indirect costs of project development on the environment and society and long-term financial commitment to post-project environmental monitoring.
5. Social barrier: the limited role of public involvement in the EIA process, and the limited attention to both direct and indirect social and cultural impacts due to project development, particularly during the monitoring stages.
6. Political barrier: limited accountability with regard to developers actually doing what they said they would with regard to managing environmental impacts, and the use of EIA as a bureaucratic and administrative exercise rather than a process to facilitate informed and responsible decision-making.

These are some of the critical challenges that currently face EIA. How these challenges are addressed, or not, will in large part determine the next phase of EIA evolution and the role of EIA as either a tool for reactive management or one that facilitates the sustainable development of the environment, the economy, and society.

KEY TERMS

biophysical environment
cost-benefit analysis
environment
environmental impact assessment
environmental impact statement

human environment
National Environmental Policy Act
scoping
screening
strategic environmental assessment

STUDY QUESTIONS AND EXERCISES

1. What are the potential benefits of conducting an environmental impact assessment?
2. How can environmental impact assessment contribute to sustainable development?
3. Obtain a small sample of EISs from your local library or government registry, or access on-line. Examine the table of contents in each and create a general list of common elements or issues addressed in each assessment. Are the contents similar in terms of the types of issues and topics addressed?
4. Why was there a sudden interest in environmental impact assessment in North America in the 1960s?
5. Given how environmental impact assessment has evolved over the years, what direction might you expect it to take in the future?
6. What requirements currently exist in your country or state for conducting EIA? When were these requirements implemented and how have they evolved over time? Is the environment defined to include both physical and human systems? Who is responsible for conducting EIA? How many assessments have been completed to date? Is there a particular sector or type of EIA that dominates? Why might this be so? Do you feel that EIA is meeting its sustainability objectives?
7. Visit some of the Web sites listed in Box 1.3 and compare the requirements and provisions for EIA to those of your own nation or state.
8. Review some of the challenges to EIA discussed in this chapter. Can you identify one or more recent EIA case studies where these challenges were evident? How might these challenges be addressed?

9. As noted in this chapter, EIA has evolved to become one of the most widely applied environmental management tools in the world. Are there certain ethical, cultural, or other elements that should be considered when applying EIA in developed versus developing nations? What elements of the process, if any, might vary from one socio-political or cultural context to the next?

10. Richard Morgan, president of the IAIA, recently raised a question as to whether EIA has passed its 'sell by' date. What do you suppose this means? Do you agree? Provide evidence to support your claim.

REFERENCES

Barrow, C.J. 1997. *Environmental and Social Impact Assessment: An Introduction*. London: Arnold.

Battelle, C. 1978. *Environmental Evaluation in Project Planning*. Columbus, Ohio: Battelle Environmental Laboratories.

Canadian Environmental Assessment Research Council (CEARC). 1988. *The Assessment of Cumulative Effects: A Research Prospectus*. Ottawa: Supply and Services Canada.

Carson, R. 1962. *Silent Spring*. Boston: Houghton Mifflin.

Council on Environmental Quality (CEQ). 1973. National Environmental Policy Act—Regulations. *Federal Register* 43: 55978–6007.

Gibson, R.B. 2002. 'From Wreck Cove to Voisey's Bay: The Evolution of Federal Environmental Assessment in Canada', *Impact Assessment and Project Appraisal* 20, 3: 151–9.

Gilpin, A. 1995. *Environmental Impact Assessment: Cutting Edge for the 21st Century*. Cambridge: Cambridge University Press.

Glasson, J., R. Therivel, and A. Chadwick. 1999. *Introduction to Environmental Impact Assessment: Principles and Procedures, Process, Practice and Prospects*, 2nd edn. London: University College London Press.

Harrop, D.O., and A.J. Nixon. 1999. *Environmental Assessment in Practice*. Routledge Environmental Management Series. London: Routledge.

IAIA and IEA. 1999. *Principles of Environmental Impact Assessment Best Practice*. Fargo, ND: IAIA.

Morgan, R. 2005. 'Have Impact Assessments Passed Their "Sell By" Date?', *Newsletter of the International Association for Impact Assessment* 16, 3: 1, 6.

Morris, P., and R. Therivel, eds. 2001. *Methods of Environmental Impact Assessment*, 2nd edn. London: Taylor and Francis Group.

Munn, R.E. 1979. *Environmental Impact Assessment: Principles and Procedures*. New York: Wiley.

Petts, J. 1999. *Handbook on Environmental Impact Assessment*. Oxford: Blackwell Science.

Rossini, F.A., and A.L. Porter. 1983. *Integrated Impact Assessment*. Social Impact Assessment Series. New York: Perseus Books.

Sadler, B. 1996. *Environmental Assessment in a Changing World: Evaluating Practice to Improve Performance*. Final report of the International Study of the Effectiveness of Environmental Assessment. Fargo, ND: IAIA.

Trudgill, S. 1990. *Barriers to a Better Environment*. London: Bellhaven Press.

Wood, C. 1995. *Environmental Impact Assessment: A Comparative Review*. London: Longman Scientific and Technical.

An Overview of Environmental Impact Assessment in Canada

EIA in Canada

Canada is recognized internationally as a nation that has contributed significantly to the development and advancement of EIA policy and practice. Since its inception in the early 1970s, thousands of EIAs have been completed in Canada. Between 2003 and 2004, for example, there were approximately 5,650 EIAs completed at the federal level alone. Most of these were completed for relatively small-scale, routine development projects and undertakings. This chapter presents an overview of Canadian EIA systems. It is not the intent here to discuss in detail Canadian EIA practice, procedure, and requirements, as these elements are addressed throughout the text. Rather, the purpose here is to outline Canadian EIA under federal, provincial, and territorial governments and land claims agreements, and to illustrate the legal development of EIA in Canada.

Canadian EIA Systems

Environmental impact assessment in Canada is part of the law of provinces, territories, Aboriginal governments, and the federal government (Figure 2.1). At the federal level, EIA is governed by the **Canadian Environmental Assessment Act** (CEAA)—the legal basis for federal EIA in Canada. The Act requires that an impact assessment be carried out when a federal authority is involved in a project. This means that a federal authority, i.e., a ministry or agency of the federal government, proposes a project; provides financial assistance for carrying out a project; transfers control or administration of federal land to enable a project to be carried out; or provides certain licences or permits enabling a project to be carried out. The purposes of the Act are:

1. to ensure that projects are considered in a careful and precautionary manner before federal authorities take action in connection with them, so that such projects do not cause significant adverse environmental effects; and
2. to encourage responsible authorities to take actions that promote sustainable development and thereby achieve or maintain a healthy environment and a healthy economy.

Figure 2.1   Provincial, Territorial, and Land Claims-based EIA Systems in Canada.   Area depicted to the north of the jurisdiction of the Northwest Territories Mackenzie Valley Resource Management Act encompasses the Inuvialuit Settlement Region, created in an agreement signed in 1984 between the Canadian government and the Inuvialuit of the western Arctic. The agreement was established to preserve Inuvialuit culture, identity, values, and regional environmental productivity. Two co-management bodies were established under the agreement to administer environmental assessment and project reviews.

Source: Map produced by J. Bronson, University of Saskatchewan.

Provincial EIA

Canada's first provincial EIA system, Ontario's Environmental Assessment Act, was enacted in 1975. All provinces and territories now have some form of EIA regulatory approval process. This includes both EIA requirements under environmental or resource management law, such as New Brunswick's Clean Environment Act and Newfoundland's Environmental Protection Act, and requirements under EIA-specific law, such as Saskatchewan's Environmental Assessment Act. To facilitate co-ordination of EIA applications and responsibilities between the government of Canada and the provinces and territories, in 1998 the Canadian Council of Ministers of the Environment signed an accord designed to improve co-operation on environmental assessment and protection. The purpose of the Canadian Council of Ministers, which consists of environment ministers from federal, provincial, and territorial governments, is to streamline the EIA process and eliminate duplication between federal and provincial or federal and territorial processes in cases where a proposed project would potentially trigger dual EIA systems. The intent of harmonization is not to transfer jurisdictional authority from the federal government to provincial or territorial governments, but rather to ensure that federal and provincial or territorial approval processes are co-ordinated so that a project proponent does not need to present the same impact statement twice or in two different formats (Meredith, 2004). Harmonization agreements on EIA have been signed between the government of Canada and the governments of Manitoba, Alberta, Saskatchewan, and British Columbia. EIA harmonization between Canada and Quebec was established in 2004 under the Canada–Quebec Agreement on Environmental Assessment Cooperation. Notwithstanding EIA harmonization, co-ordination between federal and provincial and territorial assessment processes and regulatory requirements, as well as inconsistencies between jurisdictional mandates over the role and function of EIA in project assessment and sustainability, remains a critical challenge to EIA effectiveness.

Northern EIA

North of 60°, for all intents and purposes, EIA falls primarily under federal jurisdiction. The exceptions are a federal-territorial agreement under the Yukon Territory Environmental and Socio-economic Assessment Act, and lands subject to regulation under numerous Aboriginal land claims and co-management boards. From the initial Mackenzie Valley Pipeline Inquiry of 1974–7 to the more recent Mackenzie Valley gas project, EIA in Canada's North has undergone a number of significant regulatory and legislative changes. In 1973, the government of Canada announced a policy that would permit northern Native groups to seek compensation in the form of a land claims agreement and to have more control over development activities on their traditional lands. The very first land claims-based EIA process was initiated shortly thereafter, in 1975, when the governments of Canada and Quebec and the Cree and Inuit of northern Quebec signed the James Bay and Northern Quebec Agreement. Several additional northern EIA agreements have since been established, including the **Nunavut Land Claims Agreement** and the **Mackenzie Valley Resource Management Act** (MVRMA).

Nunavut Land Claims Agreement. The signing of the 1993 Nunavut Land Claims Agreement represents by far Canada's largest Aboriginal land claims settlement and land claims-based EIA process. The Nunavut Land Claims Agreement provides the Inuit with self-government title to approximately 350,000 square kilometres of land in what had been the eastern half of the Northwest Territories. The negotiated settlement area included mineral rights for approximately 35,000 square kilometres of this land. The Agreement also included an undertaking for the implementation of a new territory, Nunavut, which was formally established in 1999 under the Agreement (Figure 2.1). Included under Article 12.2.2 of the Agreement is the establishment of an environmental assessment review board, the **Nunavut Impact Review Board**

(NIRB), which is the primary authority responsible for EIA activities in the land claims area. The mandate of the NIRB includes:

1. screening proposals to determine whether an environmental review is required;
2. gauging and determining the extent of the regional impacts of proposed projects;
3. reviewing ecosystemic and socio-economic impacts of proposed projects;
4. determining whether proposed projects should proceed and under what conditions;
5. monitoring development projects (see www.ainc-inac.gc.ca/).

The overall purpose of the NIRB is to protect and promote the future well-being of residents and communities of the Nunavut settlement area, including its ecosystem integrity.

   Mackenzie Valley Resource Management Act. For the purposes of the MVRMA, the Mackenzie Valley is defined to include all of the Northwest Territories, with the exception of Wood Buffalo National Park and the Inuvialuit Settlement Region (Figure 2.1). Proclaimed in 1998, the MVRMA was implemented by the federal government with the intent of giving northerners increased decision-making authority over natural resource and environmental development matters. A co-management board was also established for the Sahtu and Gwich'in settlement areas of the Northwest Territories to delegate responsibilities for land-use planning and for issuing land-use permits and water-use licences. As part of this establishment, a Valley-wide public board, the **Mackenzie Valley Environmental Impact Review Board** (MVEIRB), was created to undertake EIAs and panel reviews within the jurisdiction of the MVRMA. The Canadian Environmental Assessment Act no longer applies in the Mackenzie Valley, except under very specific situations where issues are transboundary in nature or deemed necessary by the Board and the federal Minister of Environment. In such cases a joint MVEIRB-CEAA review panel is established. The MVEIRB is responsible for:

1. conducting environmental impact assessments;
2. conducting environmental reviews;
3. maintaining a public registry of environmental assessments;
4. making recommendations to the Minister of the Department of Indian and Northern Affairs regarding the approval or rejection of development projects within the region.

LEGAL DEVELOPMENT OF EIA

   Environmental assessment over the last 30 years has moved towards being earlier in planning, more open and participative, more comprehensive, more mandatory, more closely monitored, more widely applied, more integrative, more ambitious, and more humble. (Gibson, 2002: 152)

The current EIA regulatory system in Canada is the product of over 30 years of legal development, legislative reform, and political controversy. EIA at the federal level is currently required by way of EIA-specific law, and it is the responsibility of the pro-ponent (the company or individual proposing the development) to undertake an EIA and prepare an EIS as per federal, territorial, and provincial regulations and require-ments. When EIA was first formally introduced to Canada in 1973, it was never the intent of the federal government that EIA would someday have a legal basis.

EARP Origins
In Canada, EIA originated with the establishment of a task force in 1970 to examine a policy and procedure for assessing the impacts of proposed developments. In 1973 the Cabinet Committee on Science, Culture and Information agreed on the need for a formal process to assess the potential impacts associated with project development, and the first Canadian EIA process, the federal **Environmental Assessment and Review Process** (EARP) Guidelines Order, was born and the **Federal Environmental Assessment Review Office** (FEARO) was created to administer its implementation. Unlike NEPA in the US, however, the Canadian EARP had no legislative basis and was therefore not legally enforceable; rather, project impact reviews were initially intended to be co-operative and voluntary, not legally binding. The result was that impact assessments under EARP were carried out inconsistently and in some cases not car-ried out at all (FEARO, 1988). Some of the earliest projects reviewed under EARP included the Point Lepreau nuclear power station (1975), eastern Arctic offshore drilling (1978), and the Arctic Pilot Project (1980). Perhaps the most significant proj-ect reviews under EARP, and ones that would change the course of EIA in Canada, were the Oldman River Dam project in Alberta (Box 2.1) and the Rafferty-Alameda Dam in Saskatchewan—both initiated without formal project assessment.

Canadian Environmental Assessment Act
Introduced to Parliament in 1992 under the Conservative government of Canada as Bill C-78, the Canadian Environmental Assessment Act was created to replace EARP. While the new Act was intended to make EIA more rigorous and systematic, it also limited the reach of EIA to only project-level decisions and not to broader planning or policy issues. The Act set out responsibilities and procedures for the environmen-tal assessment of projects involving federal authorities and established a process to ensure that impact assessment was applied early in the planning stages of proposed project developments. When the Act applies to a proposed project, the relevant fed-eral government department is designated as the 'responsible authority' and must ensure that EIA unfolds according to the principles of the Act.

   In 1994 the **Canadian Environmental Assessment Agency** was created to oversee the Act and replaced the former FEARO. The Act was proclaimed and came into force in 1995. The Act itself has several main objectives:

- to ensure that the environmental effects of projects are reviewed before federal authorities take action so that potentially significant adverse impacts can be ameliorated;
- to encourage federal authorities to take actions that promote sustainable development;

Box 2.1   Reforming EIA: Oldman River Dam Project, Alberta

Throughout the 1960s and into the 1970s, agriculture was of growing importance in Alberta. Periodic droughts, characteristic of the region's semi-arid continental climate, meant considerable uncertainties in crop production and sustainability of local water supplies. Combined, agricultural growth and climatic unpredictability reinforced earlier initiatives for the development of large-scale irrigation systems. In 1976, plans were announced to construct an earth- and rock-filled dam at the Three Rivers site in Alberta (Oldman, Castle, and Crowsnest rivers). The proposed dam was to be approximately 75 metres in height and over 3,000 metres long, with a reservoir capable of storing nearly 500 million cubic metres of water (Oldman River Dam Project, 1992). Several sites for the dam were considered, but the Three Rivers site was favoured from the outset, notwithstanding studies that identified potentially significant adverse effects on fish and wildlife, as well as impacts on ranchers whose properties would be partially divided and flooded.

The controversy over the proposed project occurred over two periods. First, between 1976 and 1980 landowners whose lands would be flooded opposed the dam, and the Peigan Indian Reserve argued that the project was intruding on lands within the Blackfoot Nation's territory. This first controversy was in part ended in 1980 when the final project site was selected. The second, and more significant, controversy emerged in 1987 when an environmental group, Friends of the Oldman River, challenged the project through the provincial courts. The environmental group was initially unsuccessful and, based on assessments under Alberta's environmental assessment process and with the approval of the federal government under the Navigable Waters Protection Act, project construction commenced in 1988. No federal assessment under the EARP Guidelines Order was initiated, notwithstanding the involvement of a federal authority. The Oldman Dam project was nearly 40 per cent complete when Friends of the Oldman River once again challenged the project in 1989, but this time in the Federal Court. The environmental group lost the court case, but in an interesting turn of events the Court of Appeals in 1992 reversed the initial decision and quashed approval for the Oldman Dam project under the Navigable Waters Protection Act. The Court of Appeals ordered that the EARP did in fact have the force of law, and the federal government was compelled to comply with the Guidelines Order. While the decision was too late to reverse any damage that had already been created by the dam project, it did pave the way for a legal requirement for EIA in Canada, which would ensure that EIA would be implemented before development takes place.

- to promote co-operation and co-ordinated action between federal and provincial governments on environmental assessments;
- to promote communication and co-operation between federal authorities and Aboriginal peoples;
- to ensure that development in Canada or on federal lands does not cause significant adverse environmental effects in areas surrounding the project;
- to ensure that there is an opportunity for public participation in the environmental assessment process (see www.ceaa.gc.ca/013/index_e.htm).

When the Act was initially created, section 72 required that a review of the Act be undertaken every five years. In 1999 Bill C-13 was introduced, An Act to Amend the

Canadian Environmental Assessment Act, and public consultations on EIA reform were organized by the Standing Committee on the Environment and Sustainable Development. Following consultation, the Standing Committee reintroduced the bill to Parliament as Bill C-9, which received royal assent on 11 June 2003, and the revised Canadian Environmental Assessment Act became law on 30 October 2003 (see www.ceaa.gc.ca/013/001/index_e.htm). Among the commitments under the new Act are the following:

1. Promotion of high-quality assessments. The revisions will potentially contribute to better decision-making in support of sustainable development by improving compliance with the Act; strengthening the role of follow-up in EIA; providing the Minister of Environment with the option of requesting follow-up studies or additional information before making a project decision; improving the consideration of cumulative effects.
2. Assessment of appropriate projects. The revised Act is intended to focus the EIA process on those projects likely to have significant adverse environmental effects, while reducing the need for and time commitment to assessing smaller, routine projects. This involves making use of a more streamlined assessment process for routine projects, exempting certain smaller projects from EIA requirements provided that specified environmental conditions are met.
3. Application of the Act fairly and consistently. EIA now extends to reserve lands, federal lands, Crown corporations, and the assessment of transboundary effects that may occur on park reserve lands and areas of land claims.
4. Improvement of co-ordination among participants. To minimize duplication, a federal environmental assessment co-ordinator has been established to facilitate EIAs that involve more than one jurisdictional authority, such as federal-provincial EIA agreements.
5. Increased certainty in the process. In addition to clarification of key terms and procedures under the Act, the process will make greater use of mediation and dispute resolution to improve EIA efficiency. The Canadian Environmental Assessment Agency is also committed to playing a greater role in building consensus and co-ordinating the dissemination of information through early involvement in the assessment process.
6. Improvement of public participation. The nature and role of public participation is to be enhanced through improved access to information, including the federal registry of impact statements and reports (www.ceaa.gc.ca/050/index_e.cfm), expanded opportunities for public input, and better incorporation of Aboriginal perspectives and values into the decision-making process.

Towards 'Higher-Order' Assessment

Perhaps the most significant advancement in Canadian EIA in recent years is the development of environmental assessment above the project level, at the level of policies, plans, and programs. In Canada, the environmental assessment of policies was included in the former EARP Guidelines Order, where 'proposal' was defined to include any initiative for which the federal government had decision-making responsibility. With the introduction of the 1992 Canadian Environmental Assessment Act, however, environmental assessment was now restricted by law to 'projects' or 'physical undertakings', and policy, plan, and program assessment was required only by way of federal cabinet directive.

In 1990 Canada announced a reform package for environmental assessment that included new legislation and a new assessment process for decisions and actions above the project level. The commitment to strategic environmental assessment was strengthened in 1999 with release of the cabinet directive on the Environmental Assessment of Policy, Plan and Program Proposals. Under the directive, an environmental assessment is required of a policy, plan, or program when a proposal is submitted to a minister or to cabinet for approval and implementation of that proposal may result in either positive or negative environmental effects. The overall purpose of this strategic approach to environmental assessment is to allow for more informed decisions in support of sustainability initiatives.

EIA DIRECTIONS

Initiated as a policy with little grip, EIA has since evolved to become a rigorous regulatory and legal framework. How this evolution has unfolded over the past 30 years, however, might be described as the dislocation of an established order in order to improve regulatory requirements, the natural transition from one system to another, or a series of unnatural breaks categorized by forces of social and legal change. However one chooses to categorize the history of EIA in Canada, Gibson (2002: 156) captures it best in suggesting that 'Canadian environmental assessment policies and laws have evolved slowly over the past 30 years. This evolution has been hesitant and uneven, though overall it has been positive.'

KEY TERMS

Canadian Environmental Assessment Act
Canadian Environmental Assessment Agency
Environmental Assessment Review Process
Federal Environmental Assessment Review
    Office

Mackenzie Valley Environmental Impact
Assessment Review Board
Mackenzie Valley Resource Management Act
Nunavut Impact Review Board
Nunavut Land Claims Agreement

STUDY QUESTIONS AND EXERCISES

1. What are some of the critical events that have shaped the course of Canadian EIA at the federal, provincial, and territorial levels?
2. When was EIA enacted in your province (or territory)? What is the name of the legislation that requires EIA? Describe some of the main features of your province's or territory's EIA legislation, such as its purpose, how it defines 'environment', and when an EIA is required. Compare this to the EIA legislation of another Canadian province or territory.
3. It has been suggested that all of Canada should be governed by a single EIA regulatory framework and assessment process. Do you agree? What might be the advantages and challenges to removing EIA responsibility and authority from provinces and territories and transferring it to a single, composite federal EIA system?

REFERENCES

Gibson, R.B. 2002. 'From Wreck Cove to Voisey's Bay: The Evolution of Federal Environmental Assessment in Canada', *Impact Assessment and Project Appraisal* 20, 3: 151–9.

Meredith, T. 2004. 'Assessing Environmental Impacts in Canada', in. B. Mitchell, ed., *Resource and Environmental Management in Canada*, 3rd edn. Toronto: Oxford University Press, 467–96.

Oldman River Dam Project. 1992. *Report of the Environmental Assessment Panel*. Report # 42. Hull, Que.: FEARO.

FOUNDATIONS OF ENVIRONMENTAL IMPACT ASSESSMENT

Nature and Classification of Environmental Impacts

ENVIRONMENTAL EFFECTS AND IMPACTS

Environmental impact assessment is about 'effects' and 'impacts'. These terms are often used interchangeably, as in section 2(1) of the Canadian Environmental Assessment Act where an environmental 'effect' is defined as any change that a project may cause in the environment, including 'any effect of any such change' on health and socio-economic conditions, on physical and cultural heritage, on the current use of lands and resources for traditional purposes by Aboriginal persons, or on any structure, site, or thing of historical, archaeological, paleontological, or architectural significance. In principle, the 'effects of any such change' on health and socio-economic conditions, for example, are in fact impacts, not effects.

A distinction can thus be made between environmental effects and environmental impacts. **Environmental effects** are changes in the condition of a particular environmental or socio-economic parameter and are usually measurable. An effect is typically defined as a process, such as soil erosion, that is set in motion by particular project actions (Figure 3.1). Actions such as road construction or dam construction do not initiate impacts per se; rather, they initiate environmental change or effects

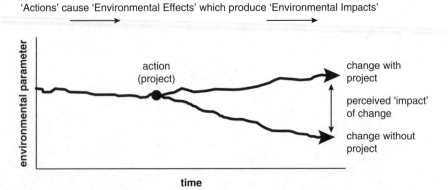

'Actions' cause 'Environmental Effects' which produce 'Environmental Impacts'

Figure 3.1 Conceptualization of an Environmental 'Impact'

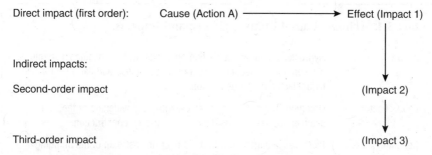

Figure 3.2 Direct, Second-Order, and Third-Order Impacts

Source: Based on Harrop and Nixon (1999).

that in turn create impacts. **Environmental impacts** are estimates or judgements of the value that society places on certain environmental effects. An impact is the net change in environmental, social, or economic well-being that results from an environmental effect, and concerns the difference between the quality of the environment with and without the proposed action. While this distinction is often said to be no more than semantics, the distinction between impacts and effects is particularly important from an environmental management perspective. For example, managing environmental 'impacts' may be no more than a damage control exercise if one does not successfully identify the project-induced environmental change (effect) that is generating the impact from the onset.

Order of Environmental Effects and Impacts

Impacts can be either **direct impacts** (first order) or **secondary impacts** (impacts resulting from the impacts of direct impacts) (Figure 3.2). The relationship among effects, first-order impacts, and second-order impacts is not always straightforward. For example, runoff from an agricultural operation may be causing enrichment of a local freshwater body with excess nutrients. The result may be excess plant growth near the shores of the water body, which may present an aesthetically pleasing waterscape. However, the subsequent eutrophication—excess plant decomposition absorbs dissolved oxygen at high rates, thereby making water unsuitable for aquatic life—generates what would be perceived as a negative secondary impact. Preston and Bedford (1988) suggest that an 'effect' is a scientific assessment of facts and an 'impact' denotes a value judgement or the relative importance of the effect (Box 3.1).

SIMPLE CLASSIFICATION OF EFFECTS AND IMPACTS

In addition to spatial and temporal scale, as introduced in Chapter 1, effects and impacts can be classified in a number of ways. In short, there is no single best set of criteria that can be used to classify all environmental effects and impacts in every project situation, but there are several common classifications of effects and impacts that should be examined for each environmental component affected by a project

---

Box 3.1　Actions, Causal Factors, Effects, and Impacts

Action	Hydroelectric dam construction on a river system. Assume a simple development project in which a river is modified and dammed for hydroelectric power generation.
Causal Factor	Dredging. The project involves dredging and widening of the river channel upstream of the dam site prior to its construction.
Condition Change/Effects	Increased sediment flow into the river downstream of the dredging site creates changes in deposition of sediments and increased turbidity.
Impacts	Direct: Decreased growth rates in fish.
	Indirect: Decline in recreational fishery.
	Several different impacts may occur as a result of the initial change in condition. For example, deposition of sediments and increased turbidity (project effects) may result in decreased growth rates in river fish (first order, direct impact), leading to a decline in the region's recreational river fishery (second order, indirect impact).

(Table 3.1). The following is by no means a comprehensive list; rather, it serves to illustrate the variety of effects and impacts that can be identified and assessed in an EIA.

Nature of Environmental Effects and Impacts

The nature of the actual effect or environmental impact includes adverse effects, incremental impacts, additive impacts, synergistic impacts, and antagonistic impacts.

Adverse effects are directly attributable to the proposed activity and are likely to cause irreversible or undesirable change, decrease the quality of the environment, or sacrifice long-term environmental integrity.

Incremental effects are marginal changes in environmental conditions that are directly attributable to the action being assessed. For example, a hydroelectric dam may lead to a slight increase in heavy metal concentrations in the river each year. It is important to consider the 'rate of change' associated with incremental effects.

Additive impacts are the consequence of separate or related actions that may be individually minor but together add up to create a significant overall impact (Figure 3.3). One might consider emissions from automobiles as additive impacts, where the emissions from a single vehicle may be not be significant but thousands of vehicles together in a crowded city, combined with the emissions from industry, create a significant overall impact on air quality. Additive impacts are not only the result of progressive increases in environmental parameters but also include, for example, the actions of several large-scale logging operations within an ecosystem. Thus, while perhaps individually manageable, together such operations present significant

Table 3.1 Classification of Environmental Effects and Impacts

Nature of effect or impact	Temporal characteristics	Magnitude of the effect or impact	Direction of change in the affected environmental parameter	Spatial extent of effect or impact	Reversibility of change of the affected environmental parameter	Probability of occurrence of an effect or impact
a) Adverse	a) Duration	a) Size	a) Increasing	a) On-site	a) Reversible	a) Likelihood
b) Incremental	b) Continuity	b) Degree	b) Decreasing	b) Off-site	b) Irreversible	b) Risk
c) Additive	c) Immediacy	c) Concentration	c) Positive			c) Uncertainty
d) Synergistic	d) Frequency		d) Negative			
e) Antagonistic	e) Regularity					

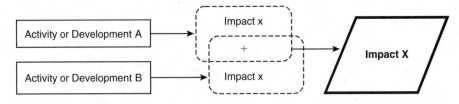

Figure 3.3　Additive Environmental Impacts

adverse additive effects on wildlife due to habitat fragmentation. Additive impacts may therefore result from two types of actions:

- Where two or more of the same type of actions are affecting the same environmental component. For example, multiple coal bed methane developments in Montana are generating cumulative effects on the landscape due to saline water discharge.
- Where a single development action produces multiple condition changes or effects that result in impacts on the same environmental component. For example, the development of a mine site may create habitat loss, changes in surface water drainage and quality, and increased noise and emissions from vehicles and operations. In combination, these can have a cumulative negative impact on wildlife.

Synergistic impacts are the result of interactions between impacts and occur when the total impacts are greater than the sum of the separate impacts (Figure 3.4). For example, a single industry located adjacent to a river system may alter water temperatures, change dissolved oxygen levels, and introduce heavy metals. Individually, each change may be tolerable for fish, but the toxicity of certain heavy metals is multiplied in high water temperatures and low dissolved oxygen content. The impact on fish due to the interaction of these effects is thus greater than the sum of the individual project-induced changes. Synergistic and additive impacts are discussed in greater detail in Chapter 11, on cumulative environmental effects.

Antagonistic impacts occur in certain situations where one adverse impact may partially cancel out another adverse impact. These are usually less common, as additional stressors to the environment more often create further disturbance and degradation. One example, however, is the reduced eutrophication of a water body

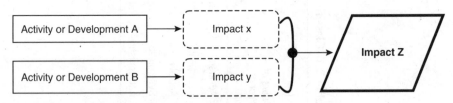

Figure 3.4　Synergistic Environmental Impacts

receiving effluents containing both chlorine and phosphates. While each is individually harmful for aquatic life, together and in moderate amounts they may be beneficial for managing eutrophication.

Temporal Characteristics of Impacts

As impacts and effects often occur over time, their characteristics may vary. Thus, for any impact or effect we can characterize its temporal nature according to the following.

Duration is the length of time that the impact occurs—for example, short-term impacts such as the noise associated with a bridge construction site or the long-term impacts associated with riverbank erosion and downstream loss of arable land.

Continuity relates to whether or not there are **continuous impacts** or effects, such as energy fields associated with transmission lines. Other impacts may be discontinuous and last only for short intervals, such as the noise from blasting at a construction site.

In regard to *immediacy*, different impacts arise at different times during the life of a particular project. Immediate impacts follow shortly after a change in the condition of the environment. For example, odour from an intensive livestock operation would occur immediately after its establishment, whereas health effects due to continued drainage from the operation into a local river system may be delayed.

The *frequency* of a change in the condition of the environment may create very different impacts. For example, local neighbourhood residents may be willing to be exposed to short-term noise due to local street maintenance activities, but they may not be willing to tolerate the construction and operation of a new airport near their homes.

Impacts that occur with *regularity*, and more predictably, may often be dealt with more easily than those that happen irregularly or that come as a surprise. For example, the 'startle effect' of low-level military flight training in Labrador on local Aboriginal populations and caribou resources is of particular concern, not because of the noise level itself but because of the surprise effect of low-flying military jets (Box 3.2).

Box 3.2   Regularity and the Startle Effect: Low-level Military Flight Training in Labrador

Canada and its NATO allies have been engaged in low-level military flight training in Labrador since the early 1980s. Approximately 6,000 to 7,000 flights are conducted annually, all below 1,000 feet above the surface. Ninety per cent of these flights are below 500 feet, and approximately 15 per cent are as low as 100 feet. Churchill Falls is the only community within the vicinity of flight training activities. However, about a dozen small Innu settlements are within the area, and these Aboriginal people rely heavily on hunting and gathering activities, particularly caribou harvesting, within the training area.

Continued

In 1990, after emerging concerns over the potential impacts associated with low-level flying activities, the Canadian Department of National Defence (DND) initiated several monitoring programs, including monitoring the potential impacts of noise on wildlife and humans. An EIA was completed in 1994 and subsequently approved in 1995 for continued flight training. A number of additional wildlife monitoring programs were implemented as a result of the EIS, particularly for the Georges River caribou herd—an important subsistence resource base for local Aboriginal communities with approximately 38,000 caribou harvested annually. Of particular concern was noise—not noise levels per se but the 'startle effect' or unpredictability of low-flying military jets.

The rate at which noise increases is referred to as the onset rate. Onset rates greater than 15 dBA per second are considered 'startling' according to DND reports. Onset rates of low-level military flights can often exceed 150 dBA per second. While the decibel rating of the noise level itself is considered tolerable, there is some concern that the unpredictability of the impact of overhead jets may affect the physiological conditions of the caribou, including reproduction performance and patterns of habitat use. While DND received considerable input from experts and local residents concerning the impacts of noise on caribou, there was no clear resolution of the actual significance of the startle effect. Both flight training and monitoring programs are ongoing.

Source: Based on CEAA (1995).

Magnitude, Direction, and Spatial Extent of Impacts

The **magnitude** of an effect refers to its size or degree, for example, the specific concentration of an emitted pollutant. That said, while many impacts can be measured in quantitative terms, such as concentration or volume, others, such as the aesthetic impact on a natural area, are subjective matters and not easily indexed or reduced to absolute quantitative terms. It is important to note that the magnitude of an impact is not necessarily related to the significance or importance of that impact. For example, removal of 5 per cent of the forest cover in an area that supports a rare or endangered species may be considered much more significant than removal of 25 per cent of forest cover in an area that supports stable native species populations.

The *direction* of change of an effect is closely associated with its magnitude. Direction of change refers to whether the affected environmental parameter is increasing or decreasing in magnitude relative to its current or previous state or **baseline condition**. For example, the construction and operation of a large-scale industry may lead to an increase in local population if it generates employment opportunities and triggers in-migration; however, it may lead to a decrease in local population if smaller industries are displaced or if quality of life in the community decreases as result of environmental change. The situation is not always this straightforward, however, as in some cases impacts are multi-directional. For example, some individuals in the affected community may see an increase in local population as a positive impact and an improvement in their quality of life, whereas others may see population increase due to industrial growth as having a negative effect on their quality of life. When characterizing environmental change, it is important to give at least some indication of direction, even if it cannot be quantified, in order to understand better the magnitude of the impact.

The *spatial extent* of an effect will vary depending on the specific parameters involved and the project environmental setting. For example, soil erosion may be highly localized and agricultural drought may be regional, while economic impacts associated with continuous drought may be national or international in extent. In the case of the Jack Pine mine project, a proposal by Shell Canada to mine bitumen deposits in the Athabasca oil sands of Alberta (see Chapter 4, Box 4.7), the spatial extent of the assessment was defined by 'local study areas' and 'regional study areas'. Local study areas were defined as those directly affected by project development. Regional study areas were identified as those from a larger geographic and ecological perspective that would experience direct and indirect impacts.

Sometimes spatial extent is classified according to **on-site impacts** and **off-site impacts**, a recognition that the impacts of any particular development or action may extend well beyond the specific project site. On-site actions may trigger environmental change, which in turn affect off-site environmental components. For example, chemical discharges at an industrial complex generate on-site environmental impacts due to soil contamination, which in turn may create off-site impacts through groundwater contamination. Such off-site impacts need not be only in terms of biophysical impacts. **Fly-in fly-out** employment arrangements, often associated with many remote mining projects, have significant impacts that extend well beyond the mine project site to the communities where the workers come from, and affect such components as family life and relationships, not to mention the effect increased air travel has on ozone depletion in the atmosphere and on demand for fossil fuels. Determining the spatial extent of EIA is a key element of the scoping process, discussed in greater detail in Chapter 6, and is critical to ensuring that all necessary impacts are considered within the scope of the assessment.

Reversibility and Irreversibility of Impacts

An additional characterization to consider when examining environmental impacts is the degree of reversibility or irreversibility. This is particularly important for determining the nature and effectiveness of impact management strategies.

An impact is considered *reversible* when it is possible to approximate the pre-disturbed condition. For example, in the case of a sanitary landfill operation the surface can be remediated to resemble its pre-project condition, or traffic congestion may return to its initial level following the end of a bridge reconstruction period. Reversing an impact, however, does not necessarily mean returning a biophysical or socio-economic environment to its *exact initial* condition, as this is neither always possible nor always desirable. Environments change and adapt irrespective of project actions, and restoring disturbed environments to their pre-disturbed condition is at best an estimate of what the affected environment might look like in the absence of development action.

Some impacts are completely *irreversible* and are of considerable importance when determining impact significance. For example, the extinction of a rare plant species during site clearing is irreversible. Other types of impacts may be reversible from a technical standpoint, but may be impractical or economically infeasible to reverse. It is common practice that many large-scale mining initiatives, for example, require drainage of a local pond or lake for tailings disposal. While it is *possible* to restore the pond or lake following mine decommissioning, it is often neither practical nor economically feasible to do so. Such impacts are often considered irreversible from any practical impact management perspective. Although considered irreversible, that is not to say that there is no need for impact management or project site cleanup. The lack of site restoration and cleanup following project operations has been particularly problematic in Canada's mining sector (Box 3.3).

Box 3.3 Reclaiming Abandoned Mine Sites in Northern Canada

The Lorado uranium mine/mill site, located approximately eight kilometres south of Uranium City in northern Saskatchewan, ceased operation in 1960. The mine site was owned by what is now EnCana West Ltd, a Calgary-based company and formerly AEC West Ltd, renamed after the 2002 merger of Pan Canadian Energy Corporation and Alberta Energy Company Ltd. The mine predated any formal requirements for EIA or post-closure site reclamation and, as a result, left behind are exposed tailings—waste rock left over after the ore is extracted—containing metals, radioactive elements, and generate acid. The Lorado mine is just one of 65 abandoned mine sites in northern Saskatchewan for which there is no corporation or government agency clearly responsible for site restoration and cleanup.

Saskatchewan's Lorado mine site is not a unique case; in fact, thousands of abandoned mine sites in northern Canada have not been reclaimed and continue to pose serious threats to the environment and to human health and safety. In the past, little could be done to ensure that mine sites were reclaimed or to prevent bankrupt companies from 'walking away' from their operations. As a result, the federal government currently spends millions of dollars each year to contain pollutants that are left behind at abandoned mine sites. In 2002, for example, the prevention of water contamination from abandoned mine sites cost Canadian taxpayers an estimated $25 million (Commissioner of Environment and Sustainable Development, 2002).

While the burden for cleanup and restoration of currently abandoned mine sites rests with the various federal and provincial governments, longer-term and viable solutions for environmental protection in Canada's mining industry now exist. Under the Canadian EIA system, and most provincial systems, mining companies are responsible for preparing mine closure plans for approval for both new and active mines and, in addition, must provide post-closure financial assurance for environmental rehabilitation or stabilization. For mining companies operating in Canada's North, a financial security deposit is required at the time of project start-up and during operation. Should a mining company claim bankruptcy the Department of Indian and Northern Affairs, the department in northern Canada responsible for land administration on behalf of the federal government, will use the security deposit to cover the eventual costs of repair, maintenance, cleanup, and closure of the mine site. If a mining company conducts the proper cleanup and site reclamation, the financial security deposit is returned.

Likelihood of Impacts

Whether an impact is considered important rests considerably on the likelihood of the impact occurring. If, for example, a particular impact is almost certain to occur, then it is more likely to be considered significant in the decision-making process than one that is determined to be highly improbable. The likelihood or probability of an impact occurring is one component used to measure **risk**—an uncertain situation involving the possibility of an undesired outcome. Risk combines the likelihood of an adverse event occurring with an analysis of the severity of the consequences associated with that event. In other words, risk can be characterized as a function of damage potential, exposure to dangerous substances such as toxic chemicals, opportunity for exposure, and the characteristics of the population at risk. For example, the elderly and young children may be much more vulnerable to exposure than other segments of the population. **Risk assessment**, then, simply refers to the process of assimilating information, identifying possible risks, risk outcomes, and the significance of those outcomes, and the likelihood and timing of their occurrence. The acceptability of a specified level of risk depends on several factors, including the catastrophic potential associated with the risk event, scientific uncertainty, distribution of risk outcomes, and understanding and familiarity of the risk. In terms of the latter, for example, an individual living on a flood plain may have significantly different perceptions of flood risk than individuals living at increasing distances from the same flood plain.

KEY TERMS

additive impacts

adverse effects

antagonistic impacts

baseline condition

continuous impacts

direct impacts

environmental effects

environmental impacts

fly-in fly-out

incremental impacts

magnitude

off-site impacts

on-site impacts

risk

risk assessment

secondary impacts

synergistic impacts

STUDY QUESTIONS AND EXERCISES

1. Consider a proposal for the construction of an industrial complex that requires forest clearing of the proposed site adjacent to a river system. Following the example presented above in Box 3.1, identify potential causal factors, condition changes, and direct and indirect impacts of the proposed action.
2. Provide an example for each of incremental, additive, and synergistic impacts.
3. Suppose a proponent submits an application for the development of a large-scale open-pit gold mine operation in your region. In small groups, brainstorm the potential 'on-site' and 'off-site' impacts. Use the 'impact characteristics' discussed in this chapter to identify those impacts you believe are most important to consider among both on-site and off-site impacts. Are there any additional criteria or characteristics that should be considered? Compare the results among groups.

REFERENCES

Canadian Environmental Assessment Agency (CEAA). 1995. *Military Flying Activities in Quebec and Labrador*. Report of the Environmental Assessment Panel. Ottawa: Minister of Supply and Services Canada.

Commissioner of the Environment and Sustainable Development. 2002. *Report of the Commissioner of the Environment and Sustainable Development*. Chapter 3: 'Abandoned Mines in the North'. Ottawa: Office of the Auditor General of Canada. Available at: <www.oag-bvg.gc.ca/domino/reports.nsf/html/c2002menu_e.html>.

Harrop, D.O., and A.J. Nixon. 1999. *Environmental Assessment in Practice*. Routledge Environmental Management Series. London: Routledge.

Preston, E., and B. Bedford. 1988. 'Evaluating Cumulative Effects on Wetland Functions: A Conceptual Overview and Generic Framework', *Environmental Management* 12, 5: 565–83.

Methods Supporting EIA Practice

METHODS AND TECHNIQUES

Literature on impact assessment often tends to use the terms 'methods' and 'techniques' in an imprecise way, treating them as synonymous. There is often confusion that 'methods', such as assessment matrices or checklists, are used to predict environmental impacts. This is not so. **Methods** are concerned with the various aspects of assessment, such as the identification and description of likely impacts and the collection and classification of data; techniques provide those data (Barrow, 1997; Canter, 1996; Bisset, 1988).

Techniques provide data that are then collated, arranged, analyzed, presented, and sometimes interpreted according to the organizational principles of the methods being used (Bisset, 1988). A technique, such as a **Gaussian dispersion model**, provides data on some parameter, such as the anticipated dispersion of air pollutants from a specific industrial development; those data are then organized according to a particular method, such as an assessment matrix, where the researcher evaluates and presents the data.

In any single EIA a number of techniques and methods may be used, but techniques and methods differ for different types of assessment. Assessment techniques, which provide the data, are much more selective than assessment methods. Many of the techniques adopted for project- and program-level assessments, such as pollutant dispersion models or population forecasting models, are not appropriate at the policy level, where the issues are by nature much broader. On the other hand, many surveying and forecasting techniques based on the use of expert judgement, such as the **Delphi technique**, are equally applicable to the policy and project levels. Methods, which are concerned with the various aspects of assessment, are applicable to all levels of assessment.

It is not possible to discuss all of the methods used in EIA in a single chapter. Thus, attention here is limited to two broad categories of EIA methods: those that support early impact and issues identification and organization; and those that facilitate communication with the public. Within each of these categories, only the most commonly used methods are explored and compared. Examples of EIA techniques are discussed at various places throughout this book.

METHODS FOR IMPACT IDENTIFICATION AND CLASSIFICATION

Methods for identifying impacts and issues are useful and systematic aids to the EIA process. Such methods alone do not constitute an EIA, as is often mistaken; rather, they serve to organize and present information for further inquiry. A variety of methods support impact and issues identification and organization of information in EIA. The choice of methods varies depending on the nature of the particular problem, the local socio-economic context, the availability of time and resources, and the goals and objectives of the assessment. No particular set of methods can do all that is required of impact assessment. Based on Bisset (1988), Canter (1992), Shopley and Fuggle (1984), and the UNEP (2002), some of the more common methods for impact identification and classification are presented here. Many of these methods are revisited in other chapters.

Examination of Similar Projects

A starting point of EIAs is often the examination of similar projects, commonly referred to as **ad hoc approaches**. Ad hoc approaches focus on identifying particular project design features, environmental impacts, and proposed management measures, as well as public reaction from other similar projects and experiences. Ad hoc approaches are valuable time-saving and cost-saving methods, particularly when multiple projects of a similar nature already exist in the proposed development area. Difficulties do emerge, however, when transferring lessons from one biophysical or socio-economic context to the next, particularly in regard to ensuring that the project environments are sufficiently similar to justify a transfer of findings. Regardless of limitations, examination of similar projects or previous relevant experience is almost always a first step when evaluating proposed developments.

Checklists

Perhaps the simplest systematic method used in EIA is the checklist. **Checklists** are comprehensive lists of environmental effects or indicators of environmental impacts designed to stimulate thinking about the possible consequences of a proposed action. They are typically used in conjunction with other methods to ensure that a prescribed list of items, regulations, or potential actions and effects is considered in the EIA and to screen particular project actions (Box 4.1).

One approach is to use a **programmed-text checklist**, similar to a questionnaire, in which a series of questions are answered. For example, a typical question for a bridge construction project might ask, 'Is there a risk of riverbank erosion?' If the answer is 'yes', then the user may be directed towards a subset of questions, such as 'Is the erosion likely to be severe enough to cause harm to fish habitat?' Programmed-text checklists are best suited for standard projects that have to be assessed frequently, such as bridge construction, municipal land-use developments, or forest access road construction. Glasson et al. (1999) propose three additional classifications of simple EIA checklists: **descriptive, questionnaire**, and **threshold of concern** (Box 4.2). The primary advantages of checklists are that they promote thinking about the range of potential issues and impacts emerging from a project in

---

Box 4.1   Partial Checklist for a Bridge Construction Project

Proposed project activities:
    dredging   ☑
    blasting   ☐
    pier construction   ☑
    traffic diversion   ☑

Affected physical components:
    water quantity   ☐
    water quality   ☑
    soil quality   ☐
    soil stability   ☑
    air quality   ☐

Affected biological components
    fish populations   ☑
    spawning habitat   ☑
    cavity nesting bird habitat   ☐
    wildlife habitat   ☐
    rare or endangered species   ☐

Affected socio-economic components
    employment   ☑
    noise   ☑
    health   ☐

---

a systematic way and are relatively efficient and easy to use. Among the disadvantages, checklists:

- are typically not comprehensive of all potential impacts, and as a result not all impacts are considered;
- are often impractical to use if they are comprehensive;
- are often too general and not tailored to specific project environments;
- make no attempt to evaluate effects either quantitatively or qualitatively;
- are subjective and qualitative, meaning that different assessors may reach different conclusions using the same checklist;
- do not consider underlying environmental systems or cause-and-effect relationships, and hence provide no conceptual understanding of impacts.

Simple Matrices

While specific design and format often vary from project to project, **matrices** are essentially two-dimensional checklists that consist of project activities on one axis and potentially affected environmental components on the other (Figure 4.1). Matrices are perhaps the most commonly used method for impact identification, and are particularly useful for identifying first-order cause-effect relationships between

Box 4.2 Types of Simple EIA Checklists

Descriptive checklists: provide guidance on how to assess certain impacts, including data requirements and potential information sources.

Data requirements	*Data or information source and techniques*
Water quality	
water uses, baseline chemicals present, runoff data	water user surveys, water quality analysis, hydrological modelling
Employment impacts	
economic base, workforce characteristics, job creation	industry survey, community profiling, regional multipliers

Questionnaire checklists: propose a set of questions that must be answered when considering the potential effects of a proposed development.

	Yes	No	?
Will the project cause pollution of air, water, or soil?			
Will there be a discharge of solid or dissolved substances to waste water?	☐	☐	☐
Is there a risk of discharge of gases that are damaging to health or environment?	☐	☐	☐
Is there risk of a potential impact on drinking water?	☐	☐	☐
Will the activity cause discharge of dust to the atmosphere?	☐	☐	☐

	Yes	No	?
Will the project cause waste problems?			
Will waste be created during operations that are hazardous to human health?	☐	☐	☐
Is there a risk that tailings may contaminate local water and soil resources?	☐	☐	☐
Have the long-term environmental impacts of mine waste been considered?	☐	☐	☐
Is the proposed management of hazardous waste in compliance with standards?	☐	☐	☐

Source: Based on NORAD (1994).

Threshold of concern checklists: list environmental components that might be affected by the project actions, specific criteria for each component, and thresholds against which the project actions can be assessed.

Component	*Criterion*	*Threshold of concern*	*Action or alternative*
Human health	noise level	maximum 12dB increase	_____
Economics	benefit-cost ratio	2:1	_____
Water supply	withdrawal rate	125,000 litres/day	

specific activities and impacts, as well as for providing a visual aid for impact summaries. The disadvantages of matrices are the same as those of checklists. EIA matrices can often be large and difficult to complete, and the volume of information can

Summary of Worst-Case Potential Impacts Prior to Mitigation							
Impact Rating – = No impact 0 = Negligible impact 1 = Minor impact 2 = Moderate impact 3 = Major impact	**Project Component**	*Physical facilities*	*Atmospheric emissions*	*Liquid and solid releases*	*Noise*	*Lights and beacons*	*Additive impacts*	*Repetitive impacts*
Environmental Components							
Marine Plants							
Phytoplankton	0	–	0	–	–	0	–
Macrophytes	–	–	–	–	–	–	–
Microbiota							
Water column	0	0	1	–	–	1	–
Sediments	0	0	1	–	–	0/1	1
Zooplankton	0	–	0/1	–	–	1	–
Ichthyoplankton	0	–	1	–	–	1	–
Macrobenthos							
Hyperbenthos	1	–	0	–	–	0	–
Epibenthos	1	–	1	–	–	1	1
Biofouling Community	1	–	1	–	–	1	1
Fish and Commercial Shellfish							
Pelagics	1	–	0	–	0	0	–
Groundfish	0	–	0	–	–	0	–
Shellfish	0	–	1	–	0	1	1

Figure 4.1 Partial Impact Identification Matrix from the Hibernia Offshore Oil Development Project

make them difficult to understand. There are several types of EIA matrices, but the most basic variations upon which more sophisticated ones are often developed include **magnitude matrices** and **interaction matrices**.

Magnitude matrices. The magnitude matrix goes beyond simple impact identification to provide a summary of impacts according to their magnitude, importance, or time frame (Glasson et al., 1999). The best-known and most comprehensive magnitude matrix is the **Leopold matrix**, originally developed for the US Geological Survey by Leopold et al. (1971). The Leopold matrix consists of a grid of 100 possible project actions along a horizontal axis and 88 environmental considerations along a vertical axis, for a total of 8,800 possible first-order project-component interactions. Each cell of the matrix consists of two values: a quantification of the magnitude of the impact and a measure of impact significance. The values within each cell typically range from −10 for a significant adverse impact to +10 for a significant beneficial impact (Box 4.3); however, more sophisticated variations are often used. While the Leopold matrix is perhaps the most comprehensive of EIA matrices, which is perhaps a drawback itself due to the sheer magnitude of the matrix, it identifies only direct project impacts, has a primarily biophysical emphasis, and does not directly provide a framework for classifying the timing or duration of impacts. Given that the Leopold matrix was developed for use on many different types of projects, it tends to generate an unwieldy amount of information for any single project application (Glasson et al., 1999). Moreover, there is no direct weighting of the affected components to reflect their relative significance.

Box 4.3 Illustration of a Typical Section of the Leopold Matrix

Matrix instructions:

1. Identify all actions across the top that are part of the proposed project.
2. Under each action, place a diagonal slash in the cell at the intersection of each component on the side of the matrix where an impact is possible.
3. Indicate the magnitude of the impact with a value from 1 to 10 in the upper left of each cell, where 1 is a low and 10 is a high magnitude. Indicate + for a positive impact or − for a negative impact. In the lower right, indicate a value from 1 to 10 for the importance of the impact.

Components and actions: modification of regime

			a) exotic flora or fauna introduction	b) biological controls	c) modification of habitat	d) alteration of ground cover	e) alteration of ground water hydrology	f) alteration of drainage	g) river control and flow modification	h) noise and vibration
A. CHEMICAL CHARACTERISTICS	1. Earth	a. mineral resources								
		b. construction material								
		c. soils								
		d. land form								
		e. force fields and radiation								
		f. unique features								
	2. Water	a. surface								
		b. ocean								
		c. underground								
		d. quality								
		e. temperature								

For example:

−10 ◄───────── Magnitude (strong negative impact)

1 ◄───────── Importance (minor, perhaps locally contained)

Source: Based on Leopold et al. (1971).

Weighted magnitude matrices. In an attempt to provide some indication of the 'relative importance' of identified project impacts, **weighted magnitude matrices** assign some measure of importance to each of the affected environmental components. In this way, values or judgements assigned to represent the potential impacts of a particular project action on an environmental component are multiplied by a weight to represent the relative importance of that component. The relative importance of the affected component may be a reflection of its current conditions, sensitivity to change, value to society, or importance in ecological functioning. An example of a simple weighted magnitude matrix is presented in Box 4.4, where individual project action-component interactions are multiplied by the component weight and summed across the rows to determine the total impact on each component. Weighted magnitude matrices are particularly useful for comparing the relative impact of project actions across environmental components or for comparing alternative project locations. Specific techniques for assigning weights to environmental components include ranking, rating, and paired comparisons. These are discussed in greater detail in Chapter 8. In the absence of a standardized system for assigning the impacts or the measures of component importance, magnitude matrices remain subjective and, like most other EIA methods, are only as good as the person assigning the values.

Box 4.4 Example of a Simple Weighted Magnitude Matrix

Affected Environmental Components		Weight (importance)	Project actions						Total impact
			blasting	side cleaning	dredging	road construction	waste disposal	equipment transport	
	air quality	0.26	−1			−1	−1		−0.78
	water quantity	0.10	−2	−3	−3				−0.80
	water quality	0.22	−2	−4	−2				−1.76*
	noise	0.04	−2		−1	−2		−2	−0.28
	habitat	0.08		−5		−3			−0.64
	wildlife	0.08	−2	−4		−2			−0.64
	human health	0.22	−2			+3	−3		−0.44

+ = positive impact   No impact =   Moderate impact = 3
− = adverse impact   Neglible impact = 1   Major impact (irreversible or long term) = 4
                            Minor impact = 2   Severe impact (permanent) = 5

*Total impact (water quality) = (0.22)(−2) + (0.22)(−4) + (0.22)(−2) = −1.76

In the above matrix the weights are distributed across the affected environmental components such that the total of all weights is '1', where the larger the weight the more important the component. In this way, all components can be given equal weight, 1/7 in the above matrix, but to increase the importance of one component requires that a trade-off be made and the importance of another component or components be decreased.

Interaction Matrices

A shortcoming of both the Leopold matrix and the weighted magnitude matrix is that neither goes beyond direct component interactions and impacts. Interaction matrices use the multiplicative properties of simple matrices to generate a quantitative impact score of the proposed project on interacting environmental components. There are two general types of interaction matrices: **component interaction matrices** and **weighted impact interaction matrices**.

First proposed by Environment Canada in 1974 (Harrop and Nixon, 1999), the objective of component interaction matrices is to identify first-, second-, and higher-order interactions and dependencies between environmental components so that indirect impacts resulting from project actions may be better understood. An example of a component interaction matrix is presented in Box 4.5. One limitation of the component interaction matrix is that for large numbers of components the data may become quite cumbersome and require computer-based assistance. Moreover, as we reveal more and more indirect linkages, we often have less and less understanding of the nature of those linkages (for example, amplifying or offsetting tendencies) and less control over their impacts. Most impact management strategies can deal effectively with only primary and secondary impacts. This raises the question of how far down the impact chain should we go when identifying project-induced impact interactions? Is there an advantage to identifying potential third-, fourth-, fifth-, or higher-order linkages and interactions? Is it practical to do so?

Similar to weighted magnitude matrices in that impacts are multiplied by the relative importance of the affected environmental components, weighted impact interaction matrices explicitly incorporate second-order or indirect impacts. One of the more common examples of weighted impact interaction matrices is the **Peterson matrix** (Peterson et al., 1974) (Box 4.6). The Peterson matrix consists of three individual component matrices: a matrix that depicts the impacts of project actions or causal factors on environmental components; a matrix depicting the impacts of the resultant environmental change on the human environment; and a vector of weights or relative importances of those human components. The initial project-environment interaction matrix is multiplied by a matrix depicting secondary human-component impacts resulting from project-induced environmental change. The result is multiplied by the relative importance of each of the human components to generate an overall impact score. The advantage of the Peterson matrix lies in the multiplicative properties of matrices and the ease of manipulation. That said, its mathematical properties are also the primary limitation of the Peterson matrix in that two negative impacts, when multiplied, generate an overall positive impact. As discussed in the previous chapter, few impacts have offsetting tendencies.

Networks

Networks serve to identify potential direct and indirect impacts that may be triggered by initial project activities (Figure 4.2). Networks are useful for identifying sequential cause-effect linkages between project actions and multiple environmental components, and can be used to describe how project activities could potentially lead to environmental changes that may affect certain components of the environment

Box 4.5   Example of a Component Interaction Matrix for an Aquatic Environment

Step 1: Initial matrix of component linkages. Where a direct link exists between those dependent components on the side axis and the supporting components across the top, a value of '1' is entered in the cell; where there is no direct link a '0' is entered. In other words, fish (row 4) are dependent on vegetation (column 1), so a value of '1' is entered in the cell. Vegetation (row 1) is not directly dependent on fish (column 4), so a value of '0' is entered.

		Supporting aquatic components				
		vegetation	plant detritus	benthic fauna	fish	avifauna
Dependent aquatic components	vegetation	0	1	0	0	0
	plant detritus	1	0	0	0	0
	benthic fauna	0	1	0	0	0
	fish	1	1	1	0	0
	avifauna	1	0	1	1	0

Step 2: Transfer all '1's' for direct impact values to a new matrix

Step 3: Multiply the initial matrix by itself for cells with '0' values only. If the result in any cell is greater than '0', then enter a value of '2' in the new matrix to indicate a 'second-order' link. If the result is zero, then enter a value of '0' in the matrix.

		Supporting aquatic components				
		vegetation	plant detritus	benthic fauna	fish	avifauna
Dependent aquatic components	vegetation	2	1	1	1	0
	plant detritus	1	2	2	2	2
	benthic fauna	2	1	0	0	0
	fish	1	1	1	2	2
	avifauna	2	2	1	1	0

The output is a component interaction matrix where a value of '1' indicates a direct link between aquatic components, and a value of '2' suggests that the components are linked indirectly through another, common component.

Step 4: To identify possible third-order linkages, the procedure is repeated by multiplying the second matrix above by the initial matrix for all cells with a value of '0' to create a new matrix. If the result for any cell is greater than '0', then enter a value of '3' to indicate third-order component linkages, and transfer all '1's' and '2's' to this new matrix. To exhaust all possible linkages, the procedure can be repeated by multiplying the third matrix by the original.

Box 4.6 Structure of the Peterson Matrix

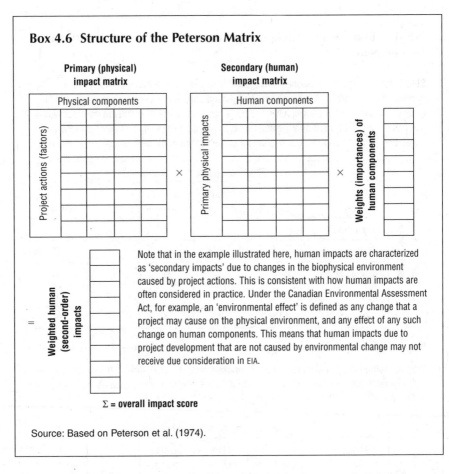

Note that in the example illustrated here, human impacts are characterized as 'secondary impacts' due to changes in the biophysical environment caused by project actions. This is consistent with how human impacts are often considered in practice. Under the Canadian Environmental Assessment Act, for example, an 'environmental effect' is defined as any change that a project may cause on the physical environment, and any effect of any such change on human components. This means that human impacts due to project development that are not caused by environmental change may not receive due consideration in EIA.

Σ = overall impact score

Source: Based on Peterson et al. (1974).

(Box 4.7). Networks can, however, become quite complicated and time-consuming as project actions and indirect linkages grow.

Perhaps the first and best-known type of network is the **Sorensen network** (Sorensen, 1971). A hybrid between a matrix and a network, the Sorensen network was initially developed to facilitate land-use planning in California (Glasson et al., 1999). A partial section of the Sorensen network is depicted in Box 4.8. The objective is the identification of direct impacts due to project actions and of subsequent cause-effect linkages so that initial changes in an environmental component can be traced to a final impact, and impact management and monitoring schemes can be designed to focus on the correct components and linkages. A primary advantage of the Sorensen network is that it presents a **holistic approach** to identifying and understanding the affected system and its components. In this way the importance of indirect effects and the interrelatedness of environmental components are recognized. The disadvantage is that there is no indication of impact magnitude, significance, or direction. Moreover, scope of the network depends on the knowledge of the network designer, and thus some important effects can be missed, particularly as the network

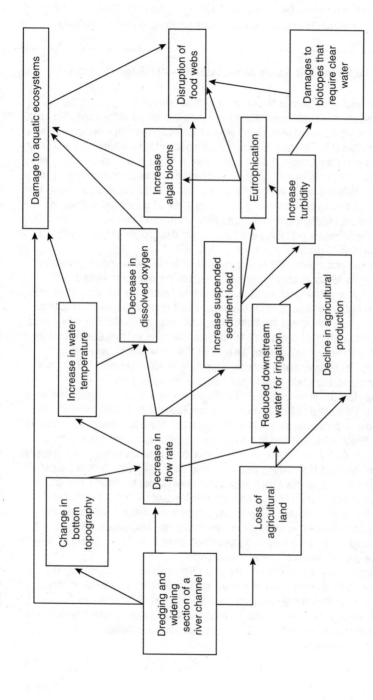

Figure 4.2 Example of a Network Diagram

grows large and complex. Inclusion of only the most important components is essential, otherwise the network becomes too large, complex, and dominated by perhaps trivial effects for which there are no direct impact management solutions.

Box 4.7 Jack Pine Oil Sands Mine, Alberta—'Linkage Diagrams'

Described by explorer Alexander MacKenzie in 1778 as 'bituminous fountains', the Athabasca oil sands of northeastern Alberta are recognized as one of the world's largest oil deposits, covering an area of over 40,000 square kilometres. The total Athabascan reserve is larger than that of Saudi Arabia. Oil sands are a mixture of sand, bitumen (83.2 per cent carbon, 10.4 per cent hydrogen, 9.4 per cent oxygen, 0.36 per cent nitrogen, 4.8 per cent sulphur), and water, from which high-quality synthetic oil can be processed. The minable oil sands in the Athabasca region are found in seams from 15 to 50 metres below the surface and are extracted using open-pit mining techniques. Open-pit mining and exploration activities have been ongoing in the Athabasca oil sands for more than 60 years, and at full production the Athabasca region is expected to supply about 10 per cent of Canada's oil needs for the next 25 years.

Shell Canada first explored the Athabasca oil sands in the 1940s, and commenced production in the 1950s. In early 2001, Shell Canada began exploration and project assessment for the Jack Pine oil sands mine (Phase 1), located approximately 70 kilometres north of Fort McMurray, Alberta, in the Clearwater Lowland physiographic division of the Saskatchewan Plain. An EIS was submitted to the Alberta Energy Utilities Board for project approval in May 2002. The proponents for the Jack Pine Mine are Shell Canada Limited, Chevron Canada Limited, and Western Oilsands Inc. The lease area for the mine covers more than 20,000 hectares and is expected to produce 31,800 cubic metres of bitumen product per calendar day, with an anticipated lifespan of over 20 years.

The major facilities of the proposed mine project include a 'truck and shovel' mining operation, a rock crusher and conveyor system, processing plant, tailings management system, and a co-generation plant, fuelled by natural gas, to produce power and steam. The purpose of Shell Canada's EIA was to examine the relationships between the activities associated with the mine development and potential impacts on the human and biophysical environment. To assist in identifying potential relationships, Shell Canada used a series of networks as 'linkage diagrams'. The linkage diagrams were used to describe how project activities would lead to environmental changes, which in turn could affect particular environmental components, and to identify key questions for impact investigation. The linkage diagrams consisted of four main components to represent project activities, potential changes in the environment, the central questions to consider, and critical links to or from other environmental or human components.

Given that the mine site itself is covered by a 1.5-metre-thick layer of muskeg and contains one open water body, a chief concern addressed in the Jack Pine EIA is effects associated with construction development and reclamation of the mine site on terrain units within the local study area. An impact linkage diagram (below) was developed to identify potential project effects, critical impact questions, and linkages to other environmental components.

Continued

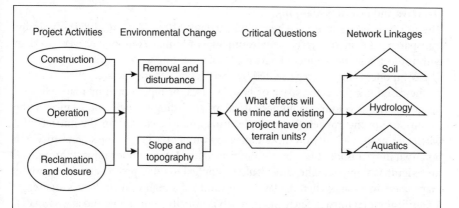

The predicted environmental impacts of the Jack Pine mine were deemed to be acceptable, generating no significant long-term impact on the environment provided that recommended mitigation measures are followed.

Source: Shell Canada (2002).

Box 4.8 Partial Section of a Sorensen Network for Forestry

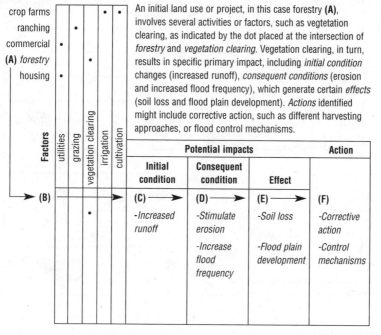

An initial land use or project, in this case forestry **(A)**, involves several activities or factors, such as vegtetation clearing, as indicated by the dot placed at the intersection of *forestry* and *vegetation clearing*. Vegetation clearing, in turn, results in specific primary impact, including *initial condition* changes (increased runoff), *consequent conditions* (erosion and increased flood frequency), which generate certain *effects* (soil loss and flood plain development). *Actions* identified might include corrective action, such as different harvesting approaches, or flood control mechanisms.

Source: Adapted from Sorensen (1971).

Overlays and Features Mapping

The use of overlays in impact mapping predates NEPA (McHarg, 1968). Typically, transparencies depicting certain environmental or human components were overlain with project activities to identify areas of potential impact or concern (Figure 4.3). The components may be represented by their actual geographic area, such as would be the case for a protected space or housing area, or represented by map cells and assigned various colours or numerical values according to their importance or the significance of the impact. For example, a map may be created for the impacts of a particular project activity on the aquatic environment, whereby the significance of a potential impact is identified by a numerical value '1' for a minor impact through '5' for a significant impact. The same could be applied to other project activities and the maps overlain so that when the layers are summed a high value indicates an area of a significant total impact. Such an approach is visually informative and allows spatial representation of impacts. Overlays and features mapping are also particularly useful when assessing land-use projects to identify potentially conflicting land-use patterns or to identify optimal locations, such as the routing of an electrical transmission line or identification of marine shipping patterns so as to identify potential conflicts with fishing activities. The approach is limited, however, to only a small number of overlays as the resultant data quickly become difficult to manage and understand.

To address this limitation, of increasing importance in EIA is the use of Geographic Information Systems, which are computer-based methods of recording, analyzing, combining, and displaying geographic information such as roads, streams, forest types, human settlement patterns, or any other feature that can be mapped on the ground. Although costly to operate and maintain, Geographic Information Systems

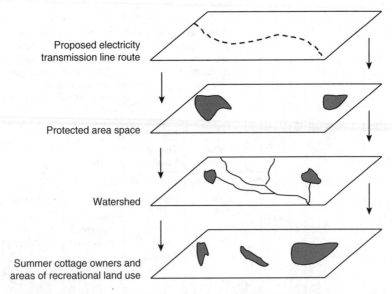

Proposed electricity
transmission line route

Protected area space

Watershed

Summer cottage owners and
areas of recreational land use

Figure 4.3 Illustration of Environmental Features Mapping for Impact Identification

are capable of handling very large data sets for each environmental parameter and, effectively, an unlimited number of parameters and data points can be processed. Some of the limitations to traditional overlay systems, however, are still inherent to modern Geographic Information Systems. The likelihood and reversibility of impacts are not easily addressed; information for large impact areas are often too coarse to be able to effectively identify intricate interactions between environmental components; a large volume of data is required, and these data are currently not available for many remote and previously unexplored areas; the approach fails to effectively capture potential issues that cannot easily be mapped; and there is difficulty in identifying non-linear impacts.

Systems Models

Models are defined as simplifications of real-world **environmental systems**; these can range from box-and-arrow diagrams to sophisticated mathematical representation based on computer simulation. Systems models based on box-and-arrow diagrams are referred to as **systems diagrams**, which consist of environmental components linked by arrows indicative of the nature of energy flow or interaction between them (Figure 4.4). Bisset (1988) explains that systems diagrams are based on the assumption that energy flows and amounts can be used as a common unit to measure project impacts on environmental components. While such diagrams are useful, they serve more of a static function and are not well-suited to understanding dynamic environmental processes.

Systems models based on mathematical representation, on the other hand, are suited for meaningful study of dynamic environmental processes. A wide range of models is available, from those that deal with single issues such as air or water quality to complex ecological system models. Examples include models to estimate the impacts of a project on air quality or simple demographic forecasting models. Perhaps the most popular, and earliest, modelling approach in EIA—**adaptive environmental assessment and management** (AEAM)—was introduced by C.S. Holling, C.J. Walters, and their associates at the University of British Columbia in the early 1970s. AEAM is an approach to simulation modelling for environmental problems that relies on teams of scientists, managers, and policy decision-makers working together to identify and define resource and environmental problems in quantifiable terms, and then model alternative impacts and management solutions. While simulation modelling was the initial focus of AEAM, the scope has became much broader in recent years and is currently applied as a broader conceptual approach to environmental assessment and management based on controlled experimentation and learning by doing (Noble, 2004a).

Expert Judgement

While not exactly a 'method' in and of itself, expert judgement does underlie many other EIA methods, from issues and impact identification to impact prediction and management. As suggested by the United Nations Environment Program (UNEP, 2002), the successful application of many EIA methods relies heavily on the nature and quality of expert judgement. Expert judgement is not new to EIA. Bailey et al.

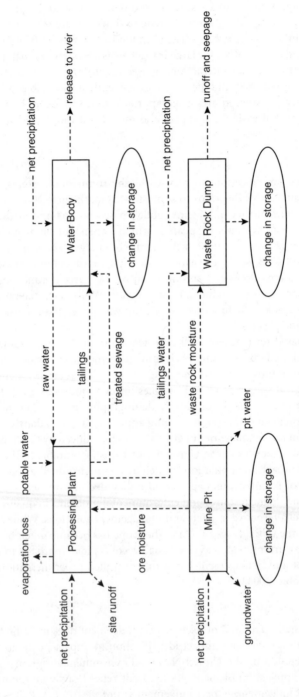

Figure 4.4 Systems Diagram Representing a Mine Site Water Balance

Source: Based on BHPB (1998).

(1992), for example, reviewed artificial waterway developments in Western Australia and found nearly two-thirds of all impact predictions to be based on knowledge and experience. As more complex methods involving integral reasoning are being used in EIA, there is, inherently, a stronger reliance on expert judgement. Although there is no standard procedure for determining expertise, Noble (2004b) identifies several guidelines that have emerged from recent EIA practices (Box 4.9).

There are three important points concerning the use of expert judgement in EIA (ibid.). First, no clear linear relationship exists between expertise and the 'quality' of impact prediction. Second, and contrary to common practice, consensus should never be the primary goal when relying on expert judgement since the same lack of knowledge that required an EIA, which relies on expert judgement in the first place, is likely to guarantee that a group of diverse experts will disagree. Any consensus that is reached may be a forced consensus and therefore highly suspect. Thus, the dissenting perspectives become of particular importance for further exploration when relying on expert judgement. Third, and in light of the limitations to consensus, emphasis should be placed on consistency when relying on expert judgement (Box 4.10). **Consistency** simply refers to a measure of the extent to which expert judgements were purposefully made or reflect random decisions; in essence, this is a measure of the degree of understanding of the problem at hand.

METHODS SELECTION

There is no single best method that can accomplish all that is required for EIA, and each method has its own strengths and limitations (Box 4.11). Horberry (1984) examined 140 EIAs and found 150 different methods being used! Multiple methods often are necessary in any single assessment. The objective is to select those methods that are best suited to the issue at hand, while minimizing the overall limitations. When selecting methods for use in impact assessment, particular attention should be given to:

- *Objectives*. What is expected to be accomplished by method application?
- *Time availability*. How much time is available to collect data and apply the method?
- *Resource availability*. What resources are available for method development and implementation, including human and financial resources?
- *Data availability*. What type and amount of data are required and available, including time series availability and availability for a range of environmental components?
- *Previous experiences*. Has the method been proven in previous applications?
- *Nature of the project*. Is the method appropriate given the size and scale of the assessment and its potential impacts?

According to Sadar (1996), desirable EIA methods should be:

- *comprehensive*—recognizing that 'environment' includes complex interrelationships between human and physical environments.

Box 4.9 Using Expert Judgement in EIA

Criteria for determining the desired number of experts and expert composition:

a) Sufficient representation of those affected by project and EIA decisions
— affected publics and interest groups
— affected sectors, government departments, and industries

b) Sufficient representation of those who affect project and EIA decisions
— public administrators
— planners and policy-makers
— scientists and researchers

c) Appropriate geographic respresentation

d) Inclusion of necessary expertise and experience

e) Practicality, given time and resources

f) Large enough number to facilitate required techniques of data analysis

g) Credibility based on the number of expert views presented

Criteria for identifying and selecting experts:

a) Experience in two or more of the specialty areas considered in the assessment

b) Current or previous management or scientific leadership role in one or more of the specialty areas considered in the assessment

c) Experience in research or administration concerning one or more of the environmental or socio-economic compents potentially affected by the project

d) Representation of a particular sector, interest, or affected geographic area

e) At least 7–10 years of combined education and experience in EIA or in one or more of the key assessment areas (disciplines) involved

f) Experience in similar types of assessments or decision processes

g) A high level of professional productivity as evidenced by academic or professional publications and research, participation in academic or industry symposia, experience in project or environmental management, or previous memebership on EIA decision or evaluation panels

h) Based on self-identified expertise—those who indicate expertise and wish to be involved

Source: Based on Noble (2004b).

- *selective*—so that they can identify those impacts that are most important and critical to project decision-making.
- *comparative*—and therefore capable of differentiating incremental impacts likely to occur in the absence of the project from project-induced change.
- *objective*—and thus able to provide unbiased measures and information.

The general rule of thumb is to use what is available, given the nature and quality of the data and the desired resolution, to get the job done.

Box 4.10 Expertise and Consistency

The notion of an expert's judgement being more consistent than that of a non-expert seems almost tautological. However, a recent study by Noble (2004b) explored the relationship between expertise and consistency based on a case study of the impacts of electricity generation. Sixty-nine individuals participated in the study and each was asked to self-identify his/her area of expertise based on the affected environmental components under consideration. Both experts and non-experts were then asked to provide their assessment of the impacts of various electricity supply alternatives on all environmental components. A consistency analysis of assessment judgements identified no significant statistical difference between the consistencies of expert versus non-expert assessments at the 95 per cent confidence interval. In fact, 60 per cent of judgements deemed to be 'extremely inconsistent' (close to random) were provided by those who self-identified themselves as experts on the particular component in question.

Box 4.11 Comparison of EIA Methods

	1	2	3	4	5	6	7	8	9
Descriptive checklists	☐	✗	✗	✗	☐	✓	✓	☐	✓
Questionnaire checklists	☐	☐	☐	☐	☐	✓	✓	☐	✓
Threshold of concern checklists	☐	☐	☐	☐	☐	✓	✓	☐	✓
Magnitude matrices (Leopold)	✗	✗	✗	✓	✓	✓	✓	✓	☐
Weighted magnitude matrices	✓	✓	✗	✓	✓	✓	✓	✓	✓
Component interaction matrices	✓	✗	✓	☐	✓	✓	✓	✗	☐
Weighted interaction matrices	✓	✓	✓	☐	✓	✓	✓	✗	☐
Network diagrams	✓	✗	✓	✗	☐	✗	✓	✗	✓
Sorensen network	✓	✗	✓	✗	☐	✗	✓	✗	☐
Overlays (GIS)	☐	✓	☐	☐	☐	✓	✗	✓	✗
Systems models	☐	☐	✓	☐	✓	✗	✗	✓	✗

1. Comprehensive of physical and human components.
2. Identifies relative importance of impacts or components.
3. Includes direct and indirect impacts.
4. Identifies prositive and negative impacts, magnitude, likelihood, and reversibility.
5. Uses qualitative and quantitative data.
6. Summarizes key impacts for further consideration.
7. Cost-effective.
8. Capable of effectively communicating a large volume of useful information.

TECHNIQUES FOR INVOLVEMENT AND COMMUNICATION

Public involvement initiatives began to develop and have an influence on environmental decision-making during the 1960s in the United States. In fact, public involvement was a critical driving force behind the initial development of EIA, and since NEPA in 1969 public involvement has increasingly been recognized as a key element in EIA practice.

There are international provisions with respect to public involvement in EIA, including the 1991 Espoo Convention on Environmental Impact Assessment in a Transboundary Context and the 1998 Aarhus Convention on Access to Information, Public Participation in Decision Making and Access to Justice in Environmental Matters. Many national EIA systems also have specific requirements for public involvement in project evaluation and decision-making, some more stringent than others. In the UK, for example, article 6 of Directive 85/337/EEC provides for public participation in EIA through opportunity for a public review of the project's EIS before the project is initiated; there is no requirement, however, for public involvement either before or during project planning. In Canada, one of the four main objectives of the Canadian Environmental Assessment Act is to ensure public participation in the EIA process. The revised Canadian Environmental Assessment Act of 2003 strengthened early public involvement in EIA during the screening phase and provided for incorporation of **traditional knowledge** during the assessment process. In practice, however, early involvement and incorporation of traditional knowledge are not required, but are at the discretion of project proponents.

Rationale for Public Involvement

It is usually recommended that public involvement commence as early as possible in the EIA process, usually during the project scoping phase. In practice, public involvement is perhaps most common during the review of the completed project EIS. However, public involvement can be beneficial to all stages of the EIA process, from initial project design to post-decision analysis and monitoring (Table 4.1).

By involving the public in the decision-making process, Mitchell (2002) suggests that it is possible to:

- define the problem more effectively;
- access a wider range of information;
- identify socially acceptable solutions;
- facilitate implementation.

While public involvement in EIA may extend the time needed during the initial project planning and scoping phases, Mitchell suggests that the initial investment is usually returned later in the process through minimizing or avoiding conflict. From a corporate perspective, Eckel et al. (1992) agree because consultation with different groups—regulators, shareholders, governments, and communities—will help to clarify expectations about environmental performance and to boost a firm's reputation with the public.

Table 4.1   Objectives of Public Involvement throughout the EIA Process

EIA stage	Public involvement objectives
Initial project design	• Early identification of affected interests and values • Identification of values relevant to site selection for conflict minimization
Screening	• Public review of decisions concerning EIA requirements • Early notification to affected interests of potential development
Scoping	• Further identification of active and inactive public • Learning about public interests and values • Elicitation of local knowledge for baseline survey • Identification of potentially significant impacts • Identification of other areas of public concern and suggestion for management • Identification of alternatives • Establishment of credibility and trust between proponent and the publics
Impact prediction and evaluation	• Elicitation of values and knowledge to assist impact prediction • Identification of criteria for project evaluation • Development of public's technical understanding of project impacts
Reporting and review	• Informing public of project details, baseline conditions, likely impacts, and proposed management measures • Obtaining public feedback on key concerns, outstanding issues, and suggestions for improved management • Identification of errors or omissions in the EIS • Providing public an opportunity to challenge EIS assumptions and predictions
Decision-making	• Resolution of potential conflicts. • Final integration of responses from EIS review
Follow-up	• Maintenance of trust and credibility • Identification of management effectiveness • Elicitation of local knowledge in data collection and monitoring change

Source: Based on Petts (1999).

Levels of Public Involvement

Judith Petts (1999: 147) defines public involvement in EIA as:

> a process of engagement, where people are enlisted into the decision process to contribute to it . . . provide for exchange of information, predictions, opinions, interests and values . . . [and] those initiating the process are open to the potential need for change and are prepared to work with different interests to develop plans or amend or even drop existing proposals.

Public involvement, while an important concern in EIA, rarely consists of highly participatory approaches where proponents are willing to alter significantly project design or implementation plans. Arnstein (1969), for example, observed that different levels of public involvement could be identified, ranging from manipulation of the public to citizen control (Box 4.12). Public involvement in EIA has focused primarily on *consulting* or *informing* the public with limited options for greater participation. In particular, public participation in EIA typically involves *providing information* to the public about the proposed project and only limited opportunity for public feedback.

Identifying the Publics

There is no such thing as 'the public' in EIA; rather there are many 'publics'—some of whom may emerge at different times during the process depending on their particular concerns and the issues involved. Mitchell (2002) makes a distinction between **active publics** and **inactive publics**. The active publics are those who affect decisions, such as industry associations, environmental organizations, and other organized interest groups. Inactive publics are those who do not typically become involved in environmental planning, decisions, or issues (Diduck, 2004). When involving the publics in EIA, it is important not to over-represent the active publics and to ensure adequate representation of the inactive publics.

Ensuring involvement of the inactive publics is a challenging task. However, a number of rather simple rules concerning notification of the publics can be undertaken to ensure that inactive publics are given fair and equal opportunity to be involved in the EIA process. For example:

- Notification should be made through media reaching the maximum number of people.
- Directly affected groups should be directly notified.
- Notification should be translated to languages of the affected publics.
- Notification should include project background information and meeting venues and identify whether critical decisions will be made.
- This information should be provided in non-technical and easy to understand language.

Box 4.12 Arnstein's Ladder of Citizen Participation

Participation	Nature of involvement	Degree of power
1. Manipulation	Rubberstamp committees	Non-participation
2. Therapy	Power holders educate or cure citizens	
3. Informing	Citizens' rights and options are identified	Degrees of tokenism
4. Consultation	Citizens are heard but not always heeded	
5. Placation	Advice is received, but not acted on	
6. Partnership	Trade-offs are negotiated	Degrees of citizen
7. Delegated power	Citizens are given management power	power
8. Citizen control	for all or parts of projects or programs	

Source: Arnstein (1969).

• Publics should be given sufficient time to prepare for participation, including access to and sufficient time for reviewing project-related documents prior to any decision being made.

Methods and Techniques

There are many different methods and techniques for public involvement and communication, and the capability and capacity of those methods and techniques vary considerably. Many different methods and techniques may be used in any single EIA at each stage of the EIA process, for different publics, and for different purposes (Table 4.2). For example, a proponent may hold information seminars early in the project design stage for those communities or interest groups most likely to be affected, whereas other communities or publics outside of the direct project impact zone may be informed at this point only through mass media information. The specific methods and techniques selected for public involvement in any EIA depend on:

• the proponent's objectives;
• the proponent's commitment to public involvement;
• the nature of legal requirements for participation;
• the sensitivity of the receiving environmental and socio-economic environment;
• the availability of time and resources;
• the magnitude of potential project impacts;
• the level of public interest.

Elements for Successful Public Involvement

While there is no recipe book for effective public involvement in EIA, based on Mitchell (1997) and Clark (1994), a number of key elements for successful public participation programs can be identified:

• *Understandable process*: The public needs to have an understanding of the EIA and decision-making process.
• *Compatibility*: Respect and trust between participants is essential to overcome potential conflicts and differences in understanding.
• *Accountability*: Proponents and decision-makers must be visibly accountable to the public.
• *Accessibility*: Proponents and decision-makers must be available to the public to address their concerns.
• *Benefits to all partners*: Real benefits of involvement must be perceived by the public.
• *Comprehensive opportunities*: All relevant interests, active and inactive, must be involved.
• *Adaptability*: There must be flexibility and willingness on the part of proponents and publics to learn.
• *Quality control*: Publics must be provided with accurate, understandable, pertinent, and timely information.

Table 4.2 Selected Techniques for Public Involvement and Communication

	Capability				Capacity		
Capability meets the criterion = ✓ *Capacity* high = ● medium = ◐ low = ○	provide information	obtain feedback	resolve conflict	identify problems and values	two-way communication	caters to special interests	number of people involved
public meetings	✓	✓		✓	◐	○	◐
public displays	✓	✓			◐	○	●
presentation to small groups	✓	✓		✓	◐	◐	○
workshops		✓	✓	✓	●	●	○
advisory committees		✓		✓	●	●	○
public review of EIS	✓	✓		✓	○	●	◐
information brochures	✓				○	◐	◐
press release inviting comments	✓	✓			○	○	●
public hearings		✓		✓	○	○	◐
task forces				✓	●	●	○
site visits	✓			✓	●	●	○
information seminars	✓	✓			●	●	○
mass media information	✓				○	○	●

Source: Based on Sadar (1996); Westman (1985).

- *Continued dialogue*: There must be ongoing communication between those responsible for project decisions and those affected by those decisions.
- *Evidence-based decisions*: Actual evidence of assimilation of public concern in the EIS and project decisions should be clear.
- *Results-oriented awareness of public participation*: Feedback to the public must be provided regarding specific decisions and the extent to which these decisions were influenced by or incorporated public input.

KEY TERMS

active publics
adaptive environmental assessment and
    management (AEAM)
ad hoc approaches
checklists
component interaction matrices

consistency
Delphi technique
descriptive checklist
environmental systems
Gaussian dispersion model
Geographic Information Systems (GIS)

holistic approach
inactive publics
interaction matrices
Leopold matrix
magnitude matrices
matrices
methods
models
networks
Peterson matrix

programmed-text checklist
questionnaire checklist
Sorensen network
systems diagrams
techniques
threshold of concern checklist
traditional knowledge
weighted interaction matrices
weighted magnitude matrices

STUDY QUESTIONS AND EXERCISES

1. In small groups, develop a simple questionnaire checklist to be used for assessing the construction phase of hydroelectric dam projects. Exchange your questionnaire checklist with other groups and critically examine other groups' checklists. Are certain items not considered in the checklist that should be included? Compile an aggregate checklist of all groups' submissions and discuss, among all groups, the advantages and limitations of a checklist approach to impact scoping.

2. Using Box 4.4 as a guide, construct, as a group, a simple matrix identifying project actions and affected environmental components for a proposed highway development. Once agreement is reached as to the project actions and affected components, break into small groups and assign weights and assessment scores to the components and impacts. Calculate the total impact score for each affected component and sum the results to generate an overall project impact score. Compare your results to those of other groups. Are there noticeable differences in the results? Why might this be so? Discuss the advantages and limitations to the magnitude matrix method. How might these limitations be addressed?

3. Use the example presented in Box 4.5 to identify potential third- and fourth-order linkages between the dependent and supporting aquatic components. Discuss the usefulness and practicality of identifying linkages beyond the first and second orders for project and impact management.

4. Complete an example of the Peterson matrix identified in Box 4.6 by adding project actions and affected components and assigning impact values on a scale from '1' to '5', where '1' indicates a minor impact and '5' indicates a major impact. Assign negative values where impacts are adverse. Assign weights to the human components on a scale of '1' to '5' where '1' indicates 'unimportant' and '5' indicates 'very important', and calculate the overall impact score. Discuss the usefulness of this value. What happened when the 'negative' impact scores were multiplied? Discuss the advantages and limitations of the Peterson matrix.

5. Sketch a simple network diagram for a proposed development that involves the clearing of forested area. Provide a statement explaining the nature of each linkage in the network. Discuss the advantages and disadvantages of the network method.

6. Obtain a completed project EIS from your local library or government registry, or access one on-line. Scan the assessment and identify the types of methods used. Compare your results to those of others. Do certain methods appear to be common to most impact statements?

7. What are the provisions for public involvement under your local or state EIA system?

8. Suppose the development of a new waste disposal facility was proposed for your area. Identify who the 'active' and 'inactive' publics might be. Based on 'Arnstein's ladder', at what level would you suggest each of these different groups be involved?

9. Identify a recent controversial development project in your region and assess how public involvement was conducted. Do you think the outcome of the project proposal or the current situation might be different if a different degree of public involvement were incorporated?

10. Discuss the advantages and disadvantages to the proponent of involving publics early in the EIA and project design process.

REFERENCES

Arnstein, S. 1969. 'A Ladder of Citizen Participation', *Journal of the American Institute of Planners* 35: 216–24.

Bailey, J., V. Hobbs, and A. Saunders. 1992. 'Environmental Auditing: Artificial Waterway Developments in Western Australia', *Journal of Environmental Management* 34: 1–13.

Barrow, C.J. 1997. *Environmental and Social Impact Assessment: An Introduction*. London: Arnold.

Broken Hill Proprieties Billiton (BHPB). 1998. *NWT Diamonds Project Environmental Impact Statement*. Vancouver: BHP Diamonds.

Bisset, R. 1988. 'Developments in EIA Methods', in P. Wathern, ed., *Environmental Impact Assessment: Theory and Practice*. London: Unwin Hyman, 47–61.

Canter, L. 1992. 'Advanced Environmental Impact Assessment Methods', paper presented at the 13th International Seminar on Environmental Assessment and Management, Centre for Environmental Management and Planning, Aberdeen.

———. 1996. *Environmental Impact Assessment*, 2nd edn. New York: McGraw-Hill.

Clark, R. 1994. 'Cumulative Effects Assessment: A Tool for Sustainable Development', *Environmental Impact Assessment Review* 12, 3: 319–22.

Diduck, A. 2004. 'Incorporating Participatory Approaches and Social Learning', in B. Mitchell, ed., *Resource and Environmental Management in Canada*, 3rd edn. Toronto: Oxford University Press, 497–527.

Eckel, L., K. Fisher, and G. Russell. 1992. 'Environmental Performance Measurement', *CMA Magazine* (Mar.): 16–23.

Glasson, J., R. Therivel, and A. Chadwick. 1999. *Introduction to Environmental Impact Assessment: Principles and Procedures, Process, Practice and Prospects*, 2nd edn. London: University College London Press.

Harrop, D.O., and A.J. Nixon. 1999. *Environmental Assessment in Practice*. Routledge Environmental Management Series. London: Routledge.

Horberry, J. 1984. 'Development Assistance and the Environment: A Question of Accountability', Ph.D. thesis, MIT.

Leopold, L.B., et al. 1971. *A Procedure for Evaluating Environmental Impact*. Washington: United States Geological Survey, Geological Survey Circular No. 645.

McHarg, I. 1968. *Design with Nature*. Garden City, NY: Natural History Press.

Mitchell, B. 1997. *Resource and Environmental Management*. Harlow, UK: Addison Wesley Longman.

———. 2002. *Resource and Environmental Management*, 2nd edn. New York: Prentice-Hall

Noble, B.F. 2004a. 'A State-of-Practice Survey of Policy, Plan, and Program Assessment in Canadian Provinces', *Environmental Impact Assessment Review* 24: 351–61.

————. 2004b. 'Applying Adaptive Environmental Management', in B. Mitchell, ed., *Resource and Environmental Management in Canada*, 3rd edn. Toronto: Oxford University Press, 442–66.

Norwegian Agency for Development Cooperation (NORAD). 1994. *Initial Environmental Assessment: Mining and Extraction of Sand and Gravel*. Oslo: NORAD.

Peterson, G.L., R.S. Gemmel, and J.L. Shofer. 1974. 'Assessment of Environmental Impacts: Multiple Disciplinary Judgments of Large-scale Projects', *Ekistics* 218: 23–30.

Petts, J. 1999. 'Public Participation and Environmental Impact Assessment', in Petts, ed., *Handbook of Environmental Impact Assessment*. London: Blackwell Science.

Sadar, H. 1996. *Environmental Impact Assessment*, 2nd edn. Ottawa: Carleton University Press.

Shell Canada. 2002. *Application for Approval of the Jack Pine Mine—Phase 1*. Environmental Impact Statement submitted to the Alberta Energy and Utilities Board and Alberta Environment. Edmonton.

Shopley, J., and R. Fuggle. 1984. 'A Comprehensive Review of Current Environmental Impact Assessment Methods and Techniques', *Journal of Environmental Management* 6: 27–42.

Sorensen, J. 1971. 'A Framework for Identification and Control of Resource Degradation and Conflict in the Multiple Use of the Coastal Zone', Department of Architecture, University of California, Berkeley.

United Nations Environment Program (UNEP). Economics and Trade Program. 2002. *Environmental Impact Assessment Training Manual*, 2nd edn. New York: UNEP.

Westman, W. 1985. *Ecology, Impact Assessment, and Environmental Planning*. New York: John Wiley.

ENVIRONMENTAL IMPACT ASSESSMENT PRACTICE

Screening Procedures

SCREENING

The number of projects that could potentially be subject to EIA is quite large; thus, the first stage in any EIA process is to determine whether the project requires an assessment and, if so, to what extent. Screening is most often the responsibility of the project proponent, but in some cases the leading regulatory EIA authority may also play a role. **Screening** simply refers to the narrowing of the application of EIA to those projects that require assessment because of perceived significant environmental effects or specific regulations. Essentially, screening is the 'trigger' for EIA and asks: 'Is an EIA required?' Screening ensures that unnecessary assessments are not carried out but that developments warranting assessment are not overlooked. However, as Barrow (1999) explains, screening caused problems during the first few years of the US NEPA. The primary screening procedure under NEPA was for 'major actions *significantly* affecting the environment,' but *significantly* was an arbitrary concept—hence the importance of criteria and approaches for project screening. The Canadian Environmental Assessment Agency defines EIA screening as a systematic approach to documenting the environmental effects of a proposed project and determining the need to eliminate or minimize the adverse effects, to modify the project plan, or to recommend further assessment through mediation or assessment by a review panel.

SCREENING APPROACHES

Is an EIA Required?
Requirements for screening are typically identified by EIA guidelines or legislation and are thus highly variable from one EIA system to the next. There are three general approaches to EIA screening: case-by-case, threshold-based, and list-based.

    Case-by-case screening involves evaluating project characteristics against a checklist of regulations and guidelines as projects are submitted. While this approach is flexible to a variety of projects and has the ability to evolve over time, it is likely to be time-consuming, inconsistent, and risks the danger of poor judgement.

    Threshold-based screening involves placing proposed projects in categories and setting thresholds for each type, such as project size, level of emissions generated, or population affected (Box 5.1). While threshold-based approaches are consistent and easy to use, a problem arises when projects lie just below the threshold. For example, if the threshold for a bridge construction project requiring an EIA is set at 50 metres,

---

Box 5.1 Threshold-based Screening Regulations in Thailand

A 1978 amendment to the Improvement and Conservation of National Environmental Quality Act provided the legislative framework for EIA in Thailand, requiring that an EIA be carried out for certain types of projects and magnitudes of projects, including but not limited to:

- Dams or reservoirs with a storage volume greater than 100 million cubic metres or storage surface area greater than 15 square kilometres.
- Irrigation projects where total irrigated area is greater than 12,800 hectares.
- Commercial ports and harbours with a capacity for vessels greater than 500 gross tonnes.
- Thermal power plants with a generation capacity greater than 10MW.

---

bridges that span 49 metres are not subject to EIA albeit they are just as likely to generate the same effects.

List-based screening involves a checklist of projects for which an EIA is (or is not) required, based on the potential of that project to generate significant effects or based on regulatory requirements. In California, for example, the California Environmental Quality Act lists projects for which a full EIA must always be completed, based on project characteristics and geographic location. As well, a negative list outlines projects for which an EIA is not required. Such lists are typically referred to as inclusive and exclusive project screening lists. **Inclusion lists** include those projects that either have mandatory or discretionary requirements for EIA. **Exclusion lists**, on the other hand, include those sorts of projects that would be subject to an EIA if they were not on the exclusion list, for example, projects carried out in response to a national emergency, projects carried out for national security reasons, or routine projects considered to be of only minor significance.

The World Bank has explicit categories of projects that require assessment (Box 5.2), but cautions that project lists should be used flexibly with reference to particular geographic setting and project scale. Many countries, including Canada, Australia, Japan, and the UK, use hybrid approaches to project screening designed to capture the advantages of each approach while minimizing the disadvantages or assessment 'loopholes'. Britain's EC Directive 85/337/EEC First Schedule, Part II 1989 Regulations, for example, identifies a list of projects for which EIA is included, some of which are defined according to certain thresholds. In 1997 the UK's Department of Environment, Transport and the Regions (DETR, 1998) proposed a threshold-based screening system with allowances for case-by-case screening (Box 5.3).

SCREENING REQUIREMENTS IN CANADA

Requirements for EIA
The Canadian Environmental Assessment Act applies only to 'projects'. This includes:

Box 5.2 World Bank Project Screening Lists

Category A Projects:

These are likely to have significant adverse environmental impacts that are diverse, unprecedented, or affect an area beyond the specific project site. A full EIA is required for such projects as:

- large-scale industrial plants
- dams and reservoirs
- port and harbour development
- large-scale irrigation
- river basin development
- hazardous or toxic materials involvement
- reclamation and new land development

- forestry and production projects
- large-scale land clearance
- oil, gas, and mineral development
- large-scale drainage or flood control
- thermal or hydropower development
- manufacture, transport, or use of pesticides
- resettlement

Category B Projects:

These are likely to have adverse impacts, but are less significant than Category A projects. Most impacts are reversible, manageable, and site-specific. Projects include:

- electricity transmission
- renewable energy development
- tourism development
- small-scale irrigation and drainage
- rural water supply or sanitation
- watershed management or rehabilitation

- agro-industries
- rural electrification
- small-scale aquaculture
- rural electricity supply
- small project maintenance or upgrading

Category C Projects:

These are likely to generate only minimal or no significant adverse environmental impacts. No EIA is required for projects concerning:

- education initiatives
- nutrition programming
- institution development
- most human resource projects

- family planning
- health initiatives
- technical assistance

Source: World Bank (1993).

- activities in relation to physical works; and
- activities not related to physical works, which are described in the Inclusion List Regulations.

If it is determined that there is in fact a project, as defined under the Act, then three additional criteria must be fulfilled before environmental assessment is applied at the federal level in Canada:

- the project must not be excluded from environmental assessment;
- a federal authority must be involved; and
- there must be a trigger to initiate the assessment.

Box 5.3   Hybrid Screening Approaches in the UK

EC Directive 85/337/EEC First Schedule, Part II of the 1989 Regulations is illustrative of *inclusion* and *threshold-based* screening, identifying projects and thresholds subject to EIA as including:

- Urban development projects that would involve a total area greater than 50 hectares for new or extended urban areas, and an area greater than two hectares within existing urban areas.
- Oil and gas pipelines that exceed 80 kilometres in length.
- Installations for the manufacture of cement.
- Industrial estate development projects that exceed 15 hectares.
- Waste-water treatment plants with a capacity greater than 10,000 population or equivalent.
- All fish and fish oil factories.
- Petroleum extraction, excluding natural gas.

In addition, there exists a *case-by-case* provision where, at the discretion of a local planning authority, an EIA may be required for projects smaller than the specified thresholds but likely to have a significant effect on the environment.

The nature of the hybrid approach is clearly illustrated by the DETR (1997) proposal, which includes an 'exclusive threshold' where projects are exempt from EIA, an 'indicative threshold' where case-by-case analyses are conducted, and an 'inclusive threshold' where EIA is always required.

EIA always required	**Inclusion threshold**
[*EIA more likely to be required, but test remains likelihood of significant environmental effect*]	
Case-by-case consideration	**Indicative threshold**
[*EIA less likely to be required, but test remains likelihood of significant environmental effect*]	
EIA not required	**Exclusion threshold** (except projects in sensitive areas)

Sources: Based on Gilpin (1999); EC Directive 85/337/EEC First Schedule Part II Regulations (1989); DETR (1998) proposed amendment to Directive 97/11.

The Inclusion List Regulations specify those activities not related to a physical work that may require an environmental assessment. These activities do not involve the construction of physical infrastructure and include the abandonment of the operations of a pipeline or mine site, the disposal or release of nuclear substances, the destruction of fish or fish habitat, activities proposed in a national park or national historic site, and projects on Aboriginal lands. In addition to EIA approval, physical activities and classes of physical activities set out under the Inclusion List Regulations typically required permitting or approval under other federal acts and regulations such as the National Parks Act, National Energy Board Act, Radiation Protection

Regulations, Navigable Waters Protection Act, Indian Act, and Migratory Bird Regulations. The official Canadian Inclusion List Regulations can be accessed on the Department of Justice Canada Web site at <http://laws.justice.gc.ca/en/index.html>.

Comprehensive Study List Regulations identify those projects likely to cause significant adverse effects or of significant public concern—for example, activities in a national park, nuclear facilities, and mining and mineral processing operations—for which an environmental assessment is required. Projects and activities for which an EIA is required under the Comprehensive Study List Regulations are typically categorized according to certain project thresholds (i.e., threshold-based screening), as discussed above (Box 5.4).

Certain projects are automatically excluded from the EIA process. This includes projects carried out in response to a national emergency and other activities described on the Exclusion List Regulations. Such projects and activities are deemed not likely to generate significant adverse effects, and in many cases they are similarly defined by screening-based thresholds. For example, the proposed expansion of an electrical transmission line by not more than 10 per cent of its current length, and that would not be carried out beyond the existing right of way, involve the release of

---

Box 5.4 Selected Screening Thresholds under the Canadian Federal Comprehensive Study List Regulations

The proposed construction, decommissioning, or abandonment of a fossil fuel-fired electrical generating station with a production capacity of 200 MW or more.

The proposed construction of an electrical transmission line with a voltage of 345 kV or more that is 75 km or more in length on a new right of way.

The proposed construction, decommissioning, or abandonment of a facility for the extraction of 200,000 m^3/year or more of groundwater or an expansion of such a facility that would result in an increase in production capacity of more than 35 per cent.

The proposed construction, decommissioning, or abandonment, or an expansion that would result in an increase in production capacity of more than 35 per cent, of an oil refinery, including a heavy oil upgrader, with an input capacity of more than 10,000 m^3/day.

The proposed construction, decommissioning, or abandonment of a heavy oil or oil sands processing facility with an oil production capacity of more than 10,000 m^3/d.

The proposed construction, decommissioning, or abandonment of a metal mine, other than a gold mine, with an ore production capacity of 3,000 tonnes/day or more.

The proposed construction, decommissioning, or abandonment, or an expansion that would result in an increase in production capacity of more than 35 per cent of a Class IA nuclear facility, i.e., a nuclear fission reactor with a production capacity of more than 25 MW (thermal).

The proposed construction of a railway line more than 32 km in length on a new right of way.

pollutants into a water body, and involve the placement of any structure in or near a water body, is excluded from assessment under the Canadian Environmental Assessment Act.

If a project is subject to an assessment under one or more of the above regulations, the federal EIA process is triggered when a federal authority has decision-making responsibility in relation to the proposed project. This may happen when a federal authority either:

- proposes a project;
- provides some form of financial assistance to assist a proponent in carrying out a specified project;
- sells, leases, or transfers control or administration of federal lands for the purposes of allowing a project to be carried out; or
- provides a licence, permit, or approval for a project that involves federal acts or regulations included in the Law List Regulations, such as the National Energy Board Act and the Canada Transportation Act, that enable a project to be carried out (see http://laws.justice.gc.ca/en/index.html).

In other circumstances, where the federal Minister of Environment receives a petition requesting a particular project to be assessed and the minister considers the project likely to cause significant adverse environmental effects across jurisdictional boundaries, then the minister has the authority to require an assessment even if the above regulatory conditions are not met.

Type of EIA Required

Not all projects must undergo the same level of assessment. For example, a proposed nuclear energy facility is likely to require a considerably larger and more complex assessment than that of a simple access road construction project. Determining the requirements for and level of EIA, however, is not always a straightforward process based solely on case-by-case, threshold, or list-based screening criteria. For example, the EIA procedures in Canada and Australia incorporate a two-tiered system for determining whether an EIA is required and, if so, what type of EIA.

Under Australia's system, a notice of intent is submitted to the Australian Environmental Protection Agency describing the nature of the project, potential environmental effects, alternatives to the project, and proposed impact management measures. This is referred to as an **initial environmental examination** (IEE) or **environmental preview report** (EPR). This information is then used to determine whether an EIA is needed, and what level of assessment is required. Based on Australia's procedures the decision stemming from an IEE may be that:

- no further assessment is required;
- a full EIA is required;
- an examination of an independent Commission of Inquiry is necessary; or
- a less comprehensive public environmental report is required.

Under the Canadian federal EIA system, once it is determined that an EIA is required, there are five assessment options as described in the Canadian Environmental Assessment Act. These include **screening EIA**, **class screening EIA**, **comprehensive study EIA**, **review panel EIA**, and **mediation EIA** (Box 5.5).

Screening EIA. A screening EIA systematically documents the potential environmental effects of a proposed project, identifies mitigation measures, and determines the need to modify the project plan or recommends further assessment to eliminate or minimize those effects. A screening EIA must address:

- the environmental effects of the project and the effects of possible accidents;
- the significance of the environmental effects;
- technically and economically feasible measures that would reduce or eliminate any significant adverse effects; and
- public comments.

If the project is likely to cause significant adverse environmental effects and it is uncertain whether these effects are justified in the circumstances, or public concerns warrant it, the project is referred to a mediator or a review panel. If a screening EIA determines that the significance of project impacts are acceptable and manageable given the circumstances, then action can be taken to allow the proposed project to proceed to the development phase. Current and archived screening EIAs can be searched at <www.ceaa.gc.ca>.

Class screening EIA. As a particular type of screening assessment, class screening EIAs are associated with routine undertakings, such as those associated with urban building projects or residential zoning. Class screening EIA helps streamline the EIA process for certain types of projects not likely to cause significant adverse environmental effects, provided that certain design standards and mitigation measures are implemented. **Model class screenings** are used to provide a generic assessment of all projects within a certain class, where information contained in a 'model' report is applied to individual projects and adapted for location-specific or project-specific information. Such screenings might include fish habitat restoration programs or the construction of forest access roads. Similarly, **replacement class screenings** provide a generic assessment of all projects within a class, but no location- or project-specific information is required and thus a screening report is not necessary for each individual project. An example would be a screening EIA prepared for special holiday events or activities in a national park or at a national historic site. Activities in a national park require assessment under the Inclusion List. They are also part of a well-defined class of activities, occur in well-understood environmental settings, are unlikely to cause significant adverse effects, do not require monitoring measures, involve straightforward planning and decision-making, and are unlikely to generate public concern. Under such cases a replacement class screening would be prepared that sufficiently addresses the nature of activities and special events in general, such as vehicle parking arrangements and management measures or the use of fireworks, but individual events need not undergo separate assessment. Examples of class screenings under the Canadian Environmental Assessment Act can be found at <www.ceaa.gc.ca/010/screenings_e.htm>.

Box 5.5 Types of Projects Subject to Different Classes of EIA in Canada

Screening EIA

Agriculture and Agri-food Canada submitted a proposal in 2002 for a screening assessment of a proposal to remediate a contaminated site in Lacombe, Saskatchewan. The project involved the operation of a mobile PCB destruction system to remove and dispose of wastes and contaminated soils.

In 2004 Parks Canada submitted a replacement class screening report for special events in the Halifax Defence Complex, an area that includes five national historic sites. No single special event or activity is defined in the report; rather, potential impacts and management measures are documented for multiple types of activities and events that would normally occur at a national historic site, including concerts, sporting events, and other community events. Issues addressed in the report concern parking, fencing, noise control, and waste management.

Aquila Networks Canada, in co-operation with Parks Canada, submitted in 2003 a model class screening report for the ongoing operations and maintenance of power distribution facilities in Banff National Park. The screening report documents the current conditions of the environment near the distribution facilities, highlights the activities involved in the operation and routine maintenance of the facilities, and outlines the potential impacts and standard environmental management practices associated with such activities.

Comprehensive Study EIA

Manitoba Hydro recently submitted a proposal to construct and operate a 200-megawatt hydroelectric generating facility, the Wuskwatim Hydro Generation Station. The proposed facility would be located on the Burntwood River in northern Manitoba. The project is still under review.

In 2002 the Minister of Environment approved the Deep Panuke offshore gas development project. PanCanadian Petroleum Limited submitted a proposal to develop the Deep Panuke offshore natural gas field, approximately 250 kilometres southeast of Halifax. The Deep Panuke gas field has estimated recoverable reserves of at least one trillion cubic feet of natural gas and 5 million barrels of liquid hydrocarbons. Based on the proposed mitigation measures, the project was considered not likely to cause significant adverse effects and public concerns did not warrant referral to a review panel or mediator.

Review Panel EIA

The Terra Nova oil field, located on the northeast Grand Banks approximately 350 km southeast of St John's, is estimated to contain 300 to 400 million barrels of recoverable oil. Petro-Canada, on behalf of its investors, submitted an application in 1996 to develop the field. A joint review panel, consisting of members from the federal Ministry of Natural Resources, Newfoundland Departments of Mines and Energy and Environment, and the Canada–Newfoundland Offshore Petroleum Board, in 1997 recommended approval of the proposal, subject to a number of recommendations.

In 1994 the Minister of Environment appointed an assessment panel to review BHP Diamond's proposal to develop and operate Canada's first diamond mine. The proposed project would involve open-pit and underground mining of five diamond-bearing kimberlite deposits at Lac de Gras, 300 kilometres northeast of Yellowknife. The review panel concluded that impacts of the project are predictable and mitigable. The project was approved in 1996, subject to certain conditions, and the panel made 29 recommendations to address environmental and socio-economic concerns.

Comprehensive study EIA. Most federal projects are assessed through screening EIA. However, projects described under the Comprehensive Study List Regulations, typically associated with large-scale, complex, and environmentally sensitive projects with greater potential for adverse environmental effects, require a comprehensive EIA. A comprehensive EIA must address the same factors as a screening EIA, as well as:

- a detailed description of the project;
- alternative means of carrying out the project that are technically and economically feasible and that consider the environmental effects of the alternative means;
- the capacity of renewable resources likely to be affected; and
- the need for and any requirements of a follow-up program.

Projects such as large-scale energy projects, mine development projects, and industrial plants typically undergo comprehensive EIAs. Such projects can also be referred to a mediator or panel review assessment if deemed appropriate by the federal Minister of Environment. For comprehensive EIAs the minister can request that a proponent submit additional information concerning project effects management and monitoring, or require that the proponent address specific public concerns prior to granting project approval. A list of active and completed comprehensive study EIAs in Canada can be found at <www.ceaa.gc.ca/010/compstudies_e.htm>.

Review panel EIA. A project assessment under review panel EIA is required if the environmental effects of a project are uncertain or potentially significant, if public concern warrants, or if the Minister of Environment states that a review by an independent panel is required for either a current screening or a comprehensive study EIA. Review panels are appointed by the federal Minister of Environment.

Mediation EIA. As a voluntary process of negotiation, a mediation EIA involves an independent mediator to help interested parties resolve their issues. Mediation EIA can be used to address all issues that arise in an EIA or it can be used in combination with an assessment by a review panel. Few mediation EIAs have been carried out in Canada.

SOME GENERAL GUIDELINES FOR PROJECT SCREENING

Barrow (1999) suggests that screening criteria for determining the need for an EIA might include, at least in principle, the following:

- Some component of the project will likely reach a specified threshold.
- The site for which development is proposed is sensitive.
- The proposed development involves known or suspected dangers or risks.
- The development potentially will contribute to cumulative impacts.
- There are unattractive input-output considerations, such as excessive labour migration or pollution output.

Based on Annex 1 of the European EIA Directive, a number of general criteria are used to determine the need for some level of EIA:

- general condition and character of the receiving environment;
- potential impacts of the proposed development;
- resilience of the affected physical and human systems to cope with potential project-induced change;
- level of confidence associated with likely project impacts;
- consistency or compatability with existing policy or planning frameworks;
- the degree of public interest or perceived effects.

Screening plays a critical role in the EIA process; it is that part of the process where a decision is made as to whether a project requires an EIA and what level of assesment is necessary. While countries with EIA systems are starting to adopt national and formal requirements and gudelines for screening, Barrow (1999) suggests that screening is not widely practised and that, in cases where screening is done, authorities often fail to screen adequately. The result is that many projects that may have significant adverse effects are often overlooked and thus proceed with little prior knowledge and understanding of their potential environmental consequences.

KEY TERMS

case-by-case screening
class screening EIA
comprehensive study EIA
environmental preview report
exclusion list
inclusion list
initial environmental examination
list-based screening

mediation EIA
model class screenings
replacement class screenings
review panel EIA
screening
screening EIA
threshold-based screening

STUDY QUESTIONS AND EXERCISES

1. What is the purpose of screening in EIA? Should *all* proposed developments be subject to EIA?
2. Discuss the relative advantages and disadvantages of case-by-case, threshold-based, and list-based screening approaches. Which type of screening approach is used in your political jurisdiction (e.g., province) ?
3. What types of projects or developments are subject to EIA in your country? Compare this to one or more of the national EIA systems listed in Table 1.3.
4. Visit the Canadian federal EIA registry at <www.ceaa.gc.ca/050/index_e.cfm>. Search the registry based on keywords such as 'mine', 'road', 'electricity' or by province or date. Identify one current or completed EIA for each of model screening, class screening, comprehensive study, and panel review. Provide a brief summary of the proposed project, including the proponent, location, and date submitted and approved—if available.

REFERENCES

Barrow, C.J. 1999. *Environmental Management: Principles and Practice*. New York: Routledge.

Department of Environment, Transport and the Regions (DETR). 1998. *Draft Town and Country Planning (Assessment of Environmental Effects) Regulations*. London: DETR.

Gilpin, A. 1999. *Environmental Impact Assessment*. Cambridge: Cambridge University Press.

World Bank. 1993. 'Sectoral Environmental Assessment', *Environmental Assessment Sourcebook Update* no. 4.

Scoping Procedures

EIA SCOPING

As discussed in Chapter 1, EIAs conducted following NEPA and in the early years of the Canadian EARP were often unfocused, full of technical details, voluminous, difficult for the public to understand, and thus of little use for timely direction for environmental management. It is now recognized that it is neither feasible nor desirable to consider *all* potential project impacts, environmental components, and issues in any single EIA. **Scoping** determines the important issues and parameters that should be addressed in an environmental assessment, establishes the boundaries of the assessment, and focuses the assessment on relevant issues and concerns. It identifies those components of the biophysical and human environment that may be affected by the project and for which there is public concern. This involves determining what elements of the project to include, what environmental components are likely to be affected, and how far removed those components are from the project.

SCOPING TYPES AND FUNCTIONS

There are two broad types of scoping: closed and open. In **closed scoping** the content and scope of the EIA are predetermined by law. Any modifications occur through closed consultations between the proponent and the competent authority or regulatory agency. In **open scoping** the content and scope of the EIA are determined by a transparent process based on consultations with various interests and publics.

No matter what the type of scoping, the scoping process serves a number of important functions in the EIA process:

- ensuring input from potential stakeholders early in the process;
- identifying public and scientific concerns and values;
- evaluating these concerns to focus the assessment and provide a coherent view of the issues;
- ensuring that key issues are identified and given an appropriate degree of attention;
- reducing the volume of unnecessarily comprehensive data and information;
- avoiding a standard inventory format to EIA that may miss key elements or issues;
- defining the spatial, temporal, and other boundaries and limits of the assessment;
- ensuring that the EIA is designed to maximize information quality for decision-making purposes.

SCOPING REQUISITES

Similar to project screening, scoping requirements and practices vary considerably from one EIA system to the next. Canada and the Netherlands, for example, have a formal scoping stage as part of a legislated EIA process. In New Zealand, there exist specific objectives for scoping processes (Box 6.1). Under UK Directive 85/337, scoping is interpreted to be in part mandatory and in part discretionary. Under the amended UK Directive 11/97, for example, the competent authority, at the request of the developer, is required to give opinions as to what should be addressed in EIA, but how scoping is carried out in terms of the consideration of alternatives or identification of indirect impacts is largely at the discretion of the proponents.

While the requirements and guidelines for scoping vary, a number of general principles should be adhered to facilitate good-practice EIA scoping:

- describe the proposed activity;
- scope project alternatives;
- establish the environmental baseline;
- establish the assessment boundaries;
- identify potential impacts and issues.

Describe the Proposed Activity

The project description serves two purposes: first, to determine the need for an environmental assessment; second, to facilitate efficiency and co-ordination of the environmental assessment. Thus, description of the proposed activity for EIA scoping must be presented at a level of detail significantly greater than at the screening stage. Under the Canadian Environmental Assessment Act three tiers of project information must be provided, including:

Box 6.1 New Zealand Ministry for the Environment Scoping Objectives

- Identify potential impacts of proposed development on the environment.
- Identify potential impacts of subsequent environmental change on society.
- Inform potentially affected people of the proposed project.
- Understand the values of individuals and groups potentially affected by the project.
- Evaluate concerns expressed and the potential environmental impacts.
- Define boundaries for any further required assessment.
- Determine the nature of any further required assessment, including analytical tools and consultation methods.
- Organize, focus, and communicate findings, potential impacts, and concerns.

Source: New Zealand Ministry for the Environment (1992).

i) General information:
   - nature, name, and proposed location of the project;
   - information on consultation with affected parties and concerned authorities;
   - details on other provincial, territorial, and federal EIA legislation or land claims agreements to which project might be subject;
   - information concerning the proponent and the nature of the industry or organization;
   - identification of affected or involved government departments and agencies, including landownership;
   - information concerning required permits and project authorizations.

ii) Project information:
   - project component structures, including size and capacities;
   - activities and activity scheduling associated with construction, operation, and decommissioning phases;
   - scheduling, engineering design specification, and land-use patterns;
   - resources and material requirements, consumption volumes, and use patterns;
   - waste production, discharges, and management.

iii) Site-specific information regarding the project:
   - project location and extent of operations;
   - location and distribution of potentially affected environmental components;
   - identification of likely affected environmental components;
   - description of previous and current land-use patterns and impacts.

Project information is normally provided by the project proponent; however, other valuable sources of project information, such as previous land-use patterns and conditions, can be obtained by consultations with the general public, professionals, governments, and special interest groups and can be based, as well, on experiences elsewhere.

'Need for' and 'purpose of' the project. Describing the 'need for' and 'purpose of' the proposed project is important for defining the project effectively. Under the Canadian Environmental Assessment Act the 'need for' a proposed project is defined as the particular problem or opportunity that the project is intending to address or satisfy (Paragraph 16(2)(a)). The 'purpose of' the project is defined as what is intended to be achieved by actually carrying out the project (Paragraph 16(1)(e)). For example, the need for a proposed electrical generation station may be defined based on the demand for electricity supply or the inability of the current system to supply reliable electricity. The purpose of the project might be to supply cost-efficient, reliable electricity to expanding city residential neighbourhoods in a manner that is profitable to the utility company. The need for a particular project and its purpose are usually established from the perspective of the project proponent and set an important context for the identification and evaluation of project alternatives.

Scope Project Alternatives

Many environmental impacts can be prevented before irrevocable project location and design decisions are made by considering options to those proposed. The consideration of alternatives is a central element to good-practice EIA, and is described by the US Council on Environmental Quality as the heart of the EIA process. Alternatives are defined as options or different courses of action to accomplish a defined end. The Canadian Environmental Assessment Act identifies two types of alternatives that should be considered—**'alternatives to'** the project and **'alternative means'** of carrying out the project (Figure 6.1).

'Alternatives to' a project are functionally different ways to meet the need and purpose of the described project. The identification of alternatives to a proposed project is at the discretion of the responsible authority for screening EIAs, and at the discretion of the Minister of Environment and/or the responsible authority for comprehensive studies. Proposing alternatives to a proposed project involves developing criteria and objectives for environmental, socio-economic, and technical variables and identifying the preferred alternative based on the relative benefits and costs of each option in reference to the identified variables.

In practice, 'alternatives to' are rarely considered comprehensively in project EIA. When they are considered they often reflect narrow agency goals or are inherently biased towards the proposed project. Alternatives to a proposed project are typically limited to the 'no action' alternative, which is interpreted as either 'no change' from the current or ongoing activity or 'no activity', which refers to not proceeding with the proposed development. This bias, to a certain extent, is understandable in that by the time a project is proposed a proponent has already dedicated considerable resources to it. For example, land for the construction of an industrial complex may already have been leased by the time a proponent submits an application for development. Many alternatives that may be more environmentally sound or socially beneficial have thus already been foreclosed by the time the EIA commences. Arguably,

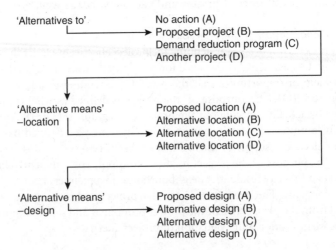

Figure 6.1 Hierarchy of Project Alternatives

'alternatives to' are best addressed at the early planning stages through the process of strategic environmental assessment (see Chapter 12), rather than by the proponent at the stage of project proposal (Box 6.2).

'Alternative means' can effectively be addressed at the project scoping stage. Alternative means refers to different options for carrying out a project—it is accepted

---

Box 6.2 Pasquia-Porcupine Forest Management Plan Alternatives Assessment

In 1995, a forest harvesting and management partnership was formed between MacMillan Bloedel Limited, one of Canada's largest forestry industries, and a subsidiary of the Saskatchewan Crown Investments Corporation, together known as the Saskfor-MacMillan Limited Partnership (SMLP). An application for development of the Pasquia-Porcupine forest was forwarded to the Saskatchewan government for approval. The Pasquia-Porcupine forest management area is located along the Saskatchewan–Manitoba provincial border within the Boreal Plain Ecozone. It encompasses an area of approximately 2 million hectares, of which more than half is suitable for commercial timber production.

The environmental assessment and forest management plan were endorsed in 1997. Consistent with the broader vision of sustainable forest management and based on early public consultation, eight plan objectives were proposed for the forest management plan: to provide quality products to meet customers' needs and provide a fair return to stakeholders;

- to provide safe and stable jobs;
- to safeguard heritage resources and traditional uses of the forest management area;
- to maintain the diversity of life forms in the area, including species and ecosystems;
- to maintain and enhance the forest ecosystem's long-term health;
- to minimize hazards from forest fires; insect infestations, and diseases;
- to protect primary resources of air, water, and soil;
- to ensure that forest areas regenerate after harvesting.

Based on these objectives, five alternatives were considered in the assessment:

1) no timber harvesting (the current baseline condition);
2) low timber priority, with reforestation based on historic levels, including retention of mature forest stands and hydrological constraints on clear-cutting;
3) intermediate timber priority, with enhanced reforestation, including retention of mature forest stands and hydrological constraints on clear-cutting;
4) high timber priority, with enhanced reforestation, reforestation of all unstocked productive land, no maintenance of old-growth forests, and no hydrological constraints on clear-cutting;
5) the option proposed by SMLP, consisting of a combination of the above, with enhanced reforestation, restoration of insufficiently restocked areas, retention of old-growth forests, and hydrological, species-specific, and ecosystem sensitivity constraints on clear-cutting.

Alternatives were assessed by a multi-disciplinary team of experts using forest land inventories and computer modelling technology. Using a Geographic Information System and a forest simulation model, changes to the forest environment were assessed for each alternative. Socio-economic impacts were similarly assessed by a panel of experts, using an input-output

Continued

model. The impacts of each alternative were assessed based on a variety of biophysical and socio-economic variables, including direct employment, income projections, tax revenues, demographic change, soil erosion, nutrient depletion, water turbidity, and long-run sustained forest growth. Option 5, the SMLP proposed option, was identified as the preferred option based on biophysical and socio-economic trade-offs.

Sources: Noble (2004); SMLP (1997).

at this stage that the proposed project is the most suitable alternative. Alternative means typically include alternative locations or alternative designs. However, it is not always possible to consider alternative locations for all proposed developments. For example, there are few possible alternative locations for a proposed gold-mining operation other than where the mineral deposit is located. That being said, alternative access roads to the mine site might avoid ecologically sensitive habitats, and alternative designs for the access road can be identified to minimize collisions with wildlife.

Alternative means should be feasible to meet the purpose of the project and can be identified based on previous or similar experiences elsewhere, expert judgement, public consultation and brainstorming, and/or through more complex decision support systems. The number of alternative means can be narrowed by eliminating those alternatives that fail to meet certain minimum project, environmental, or socio-economic requirements. For example, site screening criteria can be used to identify various environmental criteria or project objectives (Box 6.3)

Candidate alternatives must be systematically compared to identify the preferred option among those that remain for detailed impact assessment. Several methods are available for evaluating alternatives (Table 6.1); the objective is to compare across

Box 6.3 Mackenzie Valley Gas Project Alternatives Assessment

In 2003 Imperial Oil Resources Ventures Limited, Shell Canada, ConocoPhillips, ExxonMobil Canada, and the Mackenzie Valley Aboriginal Pipeline Limited Partnership submitted a preliminary information package describing the proposed Mackenzie Valley Gas Project and its related environmental and socio-economic issues.

The proposed Mackenzie Gas Project is one of the largest energy development projects in North America. The project involves the development of natural gas reserves on the Mackenzie River Delta in the Northwest Territories, natural gas processing, and the construction of a pipeline for natural gas shipment through the Mackenzie Valley to northern Alberta. At an estimated construction cost of $7 billion, the proposed pipeline will be approximately 1,220 kilometres long and transport 34 million cubic metres of natural gas per day. Over 8,000 workers will be employed at peak construction, with an estimated 150 persons employed to operate the pipeline system. Up to 32 communities will be affected by the project, including 26 in the Northwest Territories and six in northwestern Alberta.

Continued

Given the project's magnitude and northern location it was subject to EIA under three jurisdictions, the Canadian Environmental Assessment Act, the Mackenzie Valley Resource Management Act, and the Western Arctic Claims Settlement Act. A joint review panel was commissioned to administer the environmental impact review. Construction of the proposed project is expected to be completed and operational by 2009. All onshore and offshore production wells in the region will be connected to the pipeline by 2027.

Various alternatives were considered in the project EIS, including alternatives to the project (functionally different ways to produce, process, and transport gas to southern markets) and alternative means (technically and economically feasible ways to implement the proposed project). With regard to 'alternatives to', the proponent argued that there are no viable alternatives to the development and transportation of Mackenzie Delta gas to southern markets that are as advanced as the proposed project. The only true 'alternative to' is the no-go alternative, the result of which, according to the proponent, would be the loss of benefits and opportunities for both the project investors and northern residents.

With regard to 'alternative means,' several options were considered, including alternative locations for facilities and infrastructure sites, route alternatives within the preferred pipeline corridor, and alternative methods for construction and reclamation. Of considerable concern, in terms of both environmental and community impacts, was the route selected for pipeline construction. Based on field survey, baseline data collection, and discussions with local communities, the preliminary pipeline route and alternative were divided into 36 segments for alternative route evaluation. Evaluation criteria for preferred route selection included route placement (reducing pipeline length); watercourse crossings (reducing the number, complexity, and width of crossings); geotechnical considerations (avoiding steep or unstable slopes); environmental impacts (land-use patterns, critical wildlife habitat); construction considerations (reducing length of steep slopes and grading required); community interests (community feedback); and cost (relative cost of route alternatives).

Alternative route assessment occurred over three phases: first, in August 2002 to scope preliminary route options based on terrain, physical limitations, and traditional land-use patterns; second, in September 2002 based on watercourse crossings to narrow and refine the preferred route options; and third, in August 2003 to confirm the preferred route mapping and to align the final route with proposed pipeline facilities and infrastructure.

Source: Imperial Oil Resources Limited (2004).

alternatives based on similar criteria or objectives. For example, alternatives can be evaluated using a simple rating or ranking procedure based on potential economic, social, or environmental impacts or contributions (Table 6.2). In cases where the selection among alternatives is not straightforward, assessment criteria may need to be weighted to determine their relative importance (see Box 8.1). In short, the environmental and socio-economic effects of alternative means, including technical and economic feasibility, should be identified and clear justification provided for the preferred project alternative.

Establish the Environmental Baseline

Determination of the **environmental baseline** involves both the present and likely future state of the environment without the proposed project or activity. Hirsch

Table 6.1 Selected Methods for Evaluating Alternatives*

Contingent ranking	Objectives hierarchies
Contingent valuation	Opportunity costs
Cost-benefit analysis	Paired comparisons
Decision trees	Performance measures
Expert systems	Queuing models
Life-cycle assessment	Social choice theory
Linear programming	Spatial decision support systems
Moment estimation methods	Utility functions
Multi-attribute analysis	Weighted scoring
Networks	Willingness to pay
Net present value	

*Many of these are equally applicable to impact prediction.

(1980) defines **baseline study** as a description of conditions existing at points in time against which subsequent changes can be detected through monitoring. The nature and volume of baseline data vary from one project to the next (Box 6.4), but usually a number of common topics, such as air quality, water quality, and employment, are addressed and subsequently examined in detail at the later stages of prediction and management in the EIA process. Four questions should be considered when characterizing the environmental baseline:

- What do we need to know about the baseline environment to make a decision on the project?
- What are the relevant background conditions that have influenced the current environment?
- What is the likely baseline condition in the future in absence of the project?
- At what scale and dimension does the baseline environment need to be examined?

Table 6.2 Sample Alternative Site Evaluation Procedure

Candidate sites	Minimize economic cost	Minimize impacts on water quality	Topographic suitability	Ease of access	Minimize community impacts	Minimize wildlife/ habitat impacts	Total score
A	3	4	5	3	2	1	18
B	3	2	2	3	2	1	13
D	4	3	3	5	4	4	23
F	2	4	2	2	2	4	16

Legend: 5 = site meets criterion in full
4 = intermediate decision
3 = site meets criterion in part
2 = intermediate decision
1 = site does not meet criterion

Box 6.4 Experiences with Collecting Baseline Data

Baseline data may require several seasons or years to quantify sufficient ranges of natural variation and directions and rates of change, yet these data are often critical to effective impact identification and management. In the case of the La Grande-2A and La Grande-1 hydroelectric power houses located on the La Grande River in Quebec, a three-year program was initiated to establish baseline environmental conditions between 1987 and 1990. However, this is perhaps an exception to conventional practice in that baseline studies are rarely done sufficiently. Time constraints in EIA usually preclude lengthy survey and data collection programs, and impact predictions typically have to rely on existing data. In frontier areas, even existing data can be limited. In the case of the Ekati Diamond Project in the NWT, for example, most biophysical impact predictions and associated mitigation measures in the EIS were based on data collected during just one field season.

Sources: Based on Mulvihill and Baker (2001); Denis (2000).

Establish Assessment Boundaries

A large volume of baseline data *could* be collected during project scoping to characterize the biophysical (Table 6.3) and human environments (Table 6.4) and a long list of techniques for collecting that data *could* be employed, including census data, historic records, land-use plans, field surveys, and sampling procedures. However, EIA

Table 6.3 Scoping the Biophysical Environment

Air
- current pollutant concentration
- pollutant dispersion
- emission levels
- emission types
- temperatures
- windspeeds and directions

Soil
- erosion rates
- moisture content
- fertility
- organic matter
- electrical conductivity
- chemical composition
- stability
- soil pollutants

Terrestrial
- level of fragmentation
- widlife populations
- vegetation cover and composition
- air- and water-borne pollutants
- levels of light pollution
- vegetation health

Water
- surface quantity
- surface water withdrawal
- groundwater quantity
- groundwater withdrawal
- chemical content
- turbidity
- stream flow
- bank stability
- levels of eutrophication
- current pollution discharges
- fish and fish habitat

Coastal zone
- water temperature
- flood frequency
- tidal activity
- sedimentation
- marine resource populations
- bank or cliff stability

Table 6.4 Scoping the Human Environment

Economics
- local and non-local employment
- labour supply
- wage levels
- skill and education level
- retail expenditures
- material and service suppliers
- regional multiplier
- tourism

Demographics
- population
- population characteristics (family size, income, ethnicity)
- settlement patterns

Health
- quality of life (actual and perceived)
- medical standards
- worker death or injury rates
- current disease transmission
- mental and physical well-being

Housing
- public and private housing
- house prices
- homelessness and housing problems
- density and crowding

Local services
- educational services
- health services
- community services (police, fire)
- transportation services and infrastructure
- financial services

Socio-cultural
- family life
- seasonality of employment
- culture and belief systems
- crime rates, substance abuse, divorce rates
- community conflict and cohesion
- traditional foodstuffs
- community perception
- gender relations

scoping is not simply about undertaking data collection for the purpose of an 'environmental study'. Rather, scoping should establish clear boundaries for spatial and temporal assessment according to the important components of the environmental baseline. This includes limiting the amount of information to be gathered in baseline studies to a more manageable level and identifying specific objectives and indicators to guide the assessment. In this way, establishing the assessment boundaries focuses baseline studies, making them more efficient and relevant to subsequent stages of the assessment process (Figure 6.2).

Valued ecosystem components. Comprehensive baseline studies can often be a waste of time and resources. Environmental baseline studies must be undertaken within the context of clearly defined scope and objectives; otherwise, too much information is acquired that is often of too little use. While it is important to ensure that all concerned or potentially affected environmental components are given consideration, attention should focus on those **valued ecosystem components** (VECs) most likely to be affected. VECs are those aspects of the environment, physical and human, considered to be important from either public or ecological perspectives. In the Voisey's Bay EIS, discussed in Chapter 1, 17 key VECs, including water, caribou, plant communities, Aboriginal land use, and family and community, were identified based on social, economic, regulatory, and technical values and concerns. In addition, the EIS identified certain VECs that were also 'pathways' for environmental effects that may lead to effects or impacts on other VECs. For example, the VEC 'water' was identified as a biophysical effects pathway that would potentially affect other VECs: freshwater and fish habitat, waterfowl, and seabirds.

To ensure that baseline studies are purposeful and not simply a compilation of

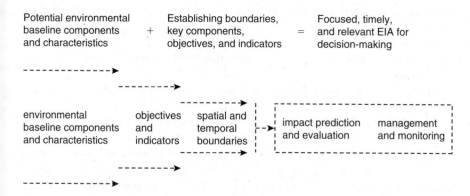

Note: There are many possible components to consider during baseline studies. Emphasizing objectives and indicators helps to focus baseline studies, generate data for impact prediction, and facilitate the design of relevant impact management and monitoring programs.

Figure 6.2 Scoping the Environmental Baseline for Focused and Informative EIA

information, it is important to identify **VEC objectives** (Table 6.5). Such objectives not only help focus baseline studies, but they are important for impact prediction and monitoring, particularly for those VECs for which considerable uncertainties exist or for which limited baseline data are available. Establishing objectives or criteria for VECs is critical to evaluating the environmental effects of the proposed project. These objectives or criteria can represent the specific parameters, guidelines, or standards

Table 6.5 Hypothetical Examples of Project VECs, Objectives, and Indicators

Component	Concern	Affected VEC	Causes	VEC targets or objectives	Indicators
surface water	lowering of lake water levels	water quantity	water withdrawal from proposed onshore oil and gas operation	ensure water levels are sufficient to maintain life support functions and existing industry and municipal uses	water volume withdrawals
wildlife	fragmentation of habitat	hunted and trapped species	proposed forestry operations	ensure habitat protection so as to enhance current wildlife populations	moose, black bear population levels

that must be met, such as a carrying capacity or set limits of environmental change, and include:

- absolute ecological or socio-economic thresholds: carrying capacity, limit of tolerance;
- acceptable limits of change: what is acceptable from a broader societal perspective;
- desired VEC conditions or objectives: desired outcomes or conditions.

Spatial bounding. What are the spatial limits of the assessment? Different VECs should be examined at different geographic scales, and for any given VEC there are various scales at which the VEC can be assessed (Box 6.5). For example, air quality can be measured across a metropolitan area or an airshed or within the confines of a particular administrative space. In other words, factors affecting air quality can be assessed across a variety of **functional scales**, ranging from point source to local, to regional (including regions of various sizes), to national (Chagnon, 1986). Thus, it is important to distinguish between two types of information: the activity information and the VEC information (Irwin and Rodes, 1992). **Activity information** characterizes the types of effects a project might generate, such as habitat fragmentation or increased use of non-renewable resources. **VEC information** refers to the processes resulting from such effects. Both of these must be taken into consideration when considering spatial boundaries. For example, the Canadian Environmental Assessment Agency's operational policy for spatial bounding in respect to offshore oil and gas exploratory drilling suggests that the spatial area of an assessment should be the 'composite' of all VEC areas (Figure 6.3), and should consider uncertainties concerning the precise location of proposed activities and effects, as well as the need to reconsider individual project bounding where multiple projects are being conducted in adjacent resource areas. Several of the methods identified in Chapter 4 can be used to assist in spatial bounding, including geographic information systems, network and

---

Box 6.5   Principles for Spatial Bounding

- Boundaries must be large enough to include relationships between the proposed project, other existing projects and activities, and the affected VECs (Cooper, 2003).
- The scope of assessment should cross jurisdictions if necessary, and allow for interconnections across systems (Shoemaker, 1994).
- Natural boundaries should be respected (Beanlands and Duinker, 1983).
- Different VECs will require assessment at different scales (Shoemaker, 1994).
- Boundaries should be set at the point where effects become insignificant by establishing a maximum detectable zone of influence (Scace et al., 2002).
- Both local and regional boundaries should be established (Canter, 1999).

Geographic boundaries for any particular assessment will vary depending on a number of factors, including the nature of the project itself, sensitivity of the receiving environment, nature of the impacts, extent of transboundary impacts, availability of baseline data, jurisdictional boundaries and co-operation, and natural physical boundaries.

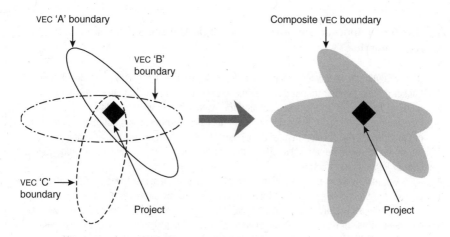

Figure 6.3 Spatial Bounding Based on Composite VEC Information

system analysis, matrices, public consultations, quantitative and physical modelling, and expert opinion.

Temporal bounding. What are the temporal limits of the assessment? The temporal scale of baseline studies should include the past, present, and future. This involves consideration of previous and current activities in the project region affecting VEC conditions, as well as potentially foreseeable activities. For example, a proposed hazardous landfill site might involve consideration of previous industrial activities that have affected the baseline environment, as well as current and proposed land-use activities. How far into the past and how far into the future will depend on the quality and quantity of data available, the certainty associated with project environmental conditions, and what the assessment is trying to achieve. Examining past conditions may be as simple as examining land-use maps or accessing census data, and it may be feasible to incorporate 50 years of historical data if deemed necessary for the proposed project. Establishing a future boundary for the assessment may be based on the end of operational life of a project, after project decommissioning and reclamation, or after affected VECs recover. Future boundaries are much more uncertain as data are often hypothetical or based on a range of future scenarios (Box 6.6). Typically, future boundaries do not exceed five years beyond project decommissioning.

Jurisdictional bounding. What are the jurisdictional limits of the assessment? Certain impacts may spread across administrative boundaries and therefore affect environments within different jurisdictions. In other cases, EIA screening may suggest that the nature of the proposed project involves both regional and national authorities. There are no general rules for establishing jurisdictional boundaries; however, it is important to consider:

- what jurisdictions are affected by the projects;
- what jurisdiction or jurisdictions have decision-making authority;
- what jurisdiction or jurisdictions are responsible for managing project impacts and over what area.

Box 6.6 Temporal Bounding in the Eagle Terrace Subdivision Assessment

Canmore, Alberta, is located in the Bow River Valley east of Banff National Park—one of Canada's most popular tourist destinations. Associated with the continued growth in tourism, however, is an increased demand for residential and visitor facilities resulting in increased pressure for housing development. In 1996 a proposal was submitted to the town of Canmore for the development of the 67-hectare Eagle Terrace residential subdivision on the slopes of the mountain valley. The proposed project would join the existing subdivisions already located near the town.

Given the involvement of a national park, the impact assessment was carried out pursuant to federal requirements. Spatial boundaries for the Eagle Terrace assessment were based on the availability of a vegetation map that covered enough of the mountain valley to include a sufficient number of actions associated with the project under review, existing land-use activities, and the information necessary to assess potential effects on wildlife VECs. The temporal boundaries of the assessment, however, were much more uncertain and based on a number of potential future scenarios, including:

- pristine scenario—a future state characteristic of conditions prior to any or extensive human development, which was simulated by removing the footprint of all developments from a GIS database;
- current scenario—assumption of the existing development, land-use, and impact conditions;
- future 'without action' scenario—assumption of future conditions predicted to occur in absence of the proposed development;
- future 'with action' scenario—assumption of future conditions that are predicted to occur with the proposed action under review.

The project EIS concluded that the proposed development (future 'with action' scenario) would incrementally contribute to the current losses of habitat, but less than 2 per cent of what had already occurred.

Source: Hegmann et al. (1999).

Identify Potential Impacts and Issues

Once the project is described and the baseline established, potential project impacts and areas of concern can be identified. The objective is to highlight for further and more intensive study those issues and impacts that will form an important part of the decision as to whether a project will be approved and under what conditions. Methods for impact identification were discussed in Chapter 4. As suggested earlier, perhaps the most common method for presenting and organizing potential impacts is an impact matrix, which identifies project activities and interactions with environmental components or VECs, and characterizes the nature of those interactions. The objective is to identify key areas of concern for further impact prediction, assessment, evaluation, and monitoring.

Several of the methods identified in Chapter 4 can be used to assist in impact identification, including geographic information systems, network and system analysis, matrices, public consultations, quantitative and physical modelling, and expert opinion. Perhaps the most common methods, however, involve a combination of expert judgement, public consultations, and scoping checklists or simple scoping matrices (Figure 6.4).

Regardless of the methods used, a critical component of impact identification is the selection of **impact indicators** (see Table 6.5). Although in certain cases the impact indicators may be the VECs themselves, impact indicators provide a measure, quantitatively or qualitatively, of the *magnitude* of an environmental impact. The EIS

Actions	Affected Components				
	wildlife	agricultural land	air	water	noise
Construction Phase					
dredging	U I 1(−) S L	U ? 1(−) S L	U ? 1(−) ? ?	C I 1(−) L R	P I 1(−) S L
site clearing	P I 3(−) P L	C I 3(−) S R	U D 1(−) L R	P D 1(−) S R	P I 1(−) S L
lake dewatering	P I 2(−) P L	U D 1(−) S L	U ? ?(?) ? ?	C I 5(−) P L	U ? ?(?) S L
road construction	C I 3(−) L R	U ? 1(+) L R	P I 1(−) S R	P D 2(−) S L	C I 1(−) S L
traffic and equipment	P I 2(−) L R	U D 1(−) S R	P I 1(−) L R	P D 1(−) S L	C I 2(−) L R

Legend:

likelihood	time frame
magnitude (direction)	
duration	extent

Likelihood: C = certain, P = probable, U = unlikely, ? = unknown
Time: I = immediate, D = delayed, ? = unknown
Magnitude: 5 = major, 3 = moderate, 1 = minor
Direction: + = positive, − = negative, ? = unknown
Duration: P = permanent, L = long term, S = short term
Extent: L = local, R = regional, N = national

Figure 6.4 Sample Project Scoping Matrix

for the Jack Pine oil sands mine in northeastern Alberta, for example, refers to such impact indicators as 'KIRs', or 'key indicator resources'. Such indicators might include specific parameters of air quality, water quality, or employment rates. Impact indicators essentially allow decision-makers to gauge environmental change efficiently by avoiding impact 'noise' and focusing on those parameters that are responsive to change, generate timely feedback, and can be traced effectively over space and time.

SCOPING CONTINUATION

A good scoping process is said to have a positive ripple effect throughout the rest of the EIA process. Scoping ensures that all of the necessary factors are considered in the assessment and at the same time sufficiently narrows the scope of the assessment to focus only on centrally important elements. While scoping is typically identified as a single phase of the EIA process, it is important to note that this is an ongoing activity in EIA and should continue to the post-project implementation and monitoring phases. Good-practice scoping is a process where the environment is continuously scanned to detect signs of adverse environmental changes or variables of importance that may not have been considered during initial project scoping exercises.

KEY TERMS

activity information
alternative means
alternatives to
baseline study
closed scoping
environmental baseline
functional scale

impact indicators
open scoping
scoping
valued ecosystem components (VECs)
VEC information
VEC objectives

STUDY QUESTIONS AND EXERCISES

1. What is the purpose of EIA scoping?
2. What are valued ecosystem components?
3. Obtain copies of topographic and political maps of your region. Suppose a new coal-fired electricity generating plant is being proposed to meet a growing demand for electricity:
   a) Identify a project location on the map sheet.
   b) Which VECs would you identify in the region as important to consider? Provide a statement as to why these environmental components are classified as VECs. In other words, what makes them so important to include in the assessment?
   c) Given current VEC conditions, what objectives would you attach to each VEC?
   d) Identify a list of indicators that you might use to assess and monitor the conditions of each VEC?
   e) What spatial boundary or boundaries would you establish for the assessment? Sketch these on a copy of the map. Identify the criteria used to determine these boundaries.
   f) What temporal boundaries might you consider?
   g) Brainstorm a number of feasible alternatives that might be considered in the project assessment, including alternative project locations.

4. Obtain a completed project EIS from your local library or government registry, or access one on-line. Identify the number and types, if any, of alternatives considered in the assessment. Were these alternatives assessed and if so to what extent? What VECs were identified? Compare your results to the findings of others.

REFERENCES

Beanlands, G.E., and P.N. Duinker. 1983. 'Lessons from a Decade of Offshore Environmental Impact Assessment', *Ocean Management* 9, 3/4: 157–75.

Canter, L. 1999. 'Cumulative effects assessment', in J. Petts, ed., *Handbook of Environmental Impact Assessment*, vol. 1, *Environmental Impact Assessment: Process, Methods and Potential*. London: Blackwell Science.

Chagnon, S. 1986. 'Atmospheric Systems: Management Perspective', in Canadian Environmental Assessment Research Council (CEARC), ed., *Cumulative Environmental Effects Assessment in Canada: From Concept to Practice*. Calgary: Alberta Society of Professional Biologists.

Cooper, L. 2003. *Draft Guidance on Cumulative Effects Assessment of Plans*. EPMG Occasional paper 03/LMC/CEA. London: Imperial College.

Denis, R. 2000. 'Lessons Derived from the Environmental Follow-up Programs on the La Grande Rivière, Downstream from the La Grande-2AQ Generating Station, James Bay, Quebec, Canada', paper presented at the International Association for Impact Assessment annual meeting, June 2000, Hong Kong.

Hegmann, G., C. Cocklin, R. Creasey, S. Dupuis, A. Kennedy, and L. Kingsley. 1999. *Cumulative Effects Assessment Practitioners Guide*. Prepared by AXYS Environmental Consulting and CEA Working Group for the Canadian Environmental Assessment Agency, Hull, Que.

Hirsch, A. 1980. 'The Baseline Study as a Tool in Environmental Impact Assessment', in *Proceedings of the Symposium on Biological Evaluation of Environmental Impacts*. Washington: Council on Environmental Quality and Fish and Wildlife Service, Department of the Interior.

Imperial Oil Resources Limited. 2004. *Environmental Impact Statement for Mackenzie Gas Project*. Ottawa: National Energy Board and Joint Review Panel.

Irwin, F., and B. Rodes. 1992. *Making Decision on Cumulative Environmental Impacts: A Conceptual Framework*. Washington: World Wildlife Fund.

Mulvihill, P., and D. Baker. 2001. 'Ambitious and Restrictive Scoping: Case Studies from Northern Canada', *Environmental Impact Assessment Review* 21: 363–84.

New Zealand Ministry for the Environment. 1992. *Guide for Scoping and Public Review Methods in Environmental Impact Assessment*. Wellington: Ministry for the Environment.

Noble, B.F. 2004. 'Integrating Strategic Environmental Assessment with Industry Planning: A Case Study of the Pasquia-Porcupine Forest Management Plan, Saskatchewan, Canada', *Environmental Management* 33, 3: 401–11.

Saskfor-MacMillan Limited Partnership (SMLP). 2004. *Twenty-year Forest Management Plan and Environmental Impact Statement for the Pasquia-Porcupine Forest Management Area*. Assessment main document. Regina: SMLP.

Scace, R., E. Grifone, and R. Usher. 2002. *Ecotourism in Canada*. Ottawa: Canadian Environmental Advisory Council.

Shoemaker, D. 1994. *Cumulative Environmental Assessment*. Waterloo, Ont.: University of Waterloo, Department of Geography.

Predicting Environmental Impacts

IMPACT PREDICTION

Impact prediction is fundamental to EIA. A prediction is a statement specifying, without direct measurement, the past, present, or future condition of a particular environmental system component, given particular system characteristics (Duinker and Baskerville, 1986). Not all aspects of project impacts are predicted, as this would be neither feasible, given time and resources, nor desirable, given the volume of information that can be effectively processed in any single EIA. Prediction involves identification of potential changes in *impact indicators* of environmental receptors (VECs) identified during EIA scoping and requires:

- determining the initial or reference state (baseline);
- predicting the future state without project development;
- predicting the future state with project development.

Predicting the baseline trend is important to understanding the true nature of the predicted impact. For example, if employment in the area for which the project is proposed has been steadily increasing over the past five years, then this baseline trend should be predicted in absence of the project in order to allow for a more informed decision concerning the predicted employment contributions of the proposed project. Similarly, if a project is predicted to have a negative effect on fish populations, then this information should be considered in light of recent and predicted trends of fish populations in the area.

Predicting the potential environmental impacts of proposed projects is a complex and often uncertain task, since many cause-effect relationships are unknown and physical and human environments are 'moving targets'. Thus, Morris and Therivel (2001) suggest that impact prediction requires several elements, including:

- sound understanding of the nature of the proposed undertaking;
- knowledge of the outcomes of similar projects;
- knowledge of past, present, or approved projects whose impacts may interact with the proposed undertaking;
- predictions of the project's impacts on other environmental and socio-economic components that may interact with those directly affected by the project;
- information about environmental and socio-economic receptors and how they might respond to change.

BASIC REQUIREMENTS OF IMPACT PREDICTIONS

Impact prediction should provide insight into the specific characteristics of potential impacts. These characteristics include:

- the nature of the predicted impact (for example, adverse, additive, antagonistic);
- temporal characteristics;
- magnitude, direction, and spatial extent;
- degree of reversibility;
- the likelihood that the predicted impact will actually occur.

These are basic principles of impact classification and are critical to determining impact significance, as discussed in the next chapter. In addition, two underlying principles essentially determine the usefulness of impact predictions for managing project impacts: accuracy and precision.

Accuracy and Precision

Impact predictions such as 'slight reduction' or 'minor effect' are of little value for understanding and managing the actual impacts of project developments. It is possible to generate quite accurate impact predictions when such predictions are couched in vague language. The value of such predictions, however, is limited in terms of avoiding or managing project impacts. **Accuracy** refers to the extent of system-wide bias in impact predictions. **Precision** refers to the level of exactness associated with impact prediction (Figure 7.1). Accuracy, then, depends on the level of

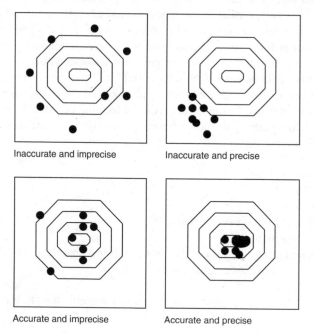

Inaccurate and imprecise   Inaccurate and precise

Accurate and imprecise   Accurate and precise

Figure 7.1 Accuracy versus Precision

precision required. One can be consistently accurate when impact predictions are imprecise. As the desired level of precision increases, then the possibility of inaccuracy in impact prediction also increases. The fact that an impact prediction is accurate does not necessarily constitute a useful impact prediction if the level of precision is relaxed. However, if more precise impact predictions are desired then accuracy may be forfeited (De Jongh, 1988).

Although prediction is the cornerstone of EIA, there is a lack of attention given to the manner in which predictions about anticipated effects are phrased. There are two dimensions to predictive accuracy: (1) the logical validity of the impact prediction; (2) the relative severity of the actual versus predicted impact. Buckley (1991) explains that in cases where actual impacts prove less severe than anticipated, predictions may still be logically incorrect if they were expressed as 'impact parameter will be equal to predicted numerical value' rather than 'impact parameter will be less than or equal to predicted numerical value.' However, whether an impact prediction is logically correct is perhaps less meaningful than the accuracy of the relative severity of actual versus predicted impacts, that is, impacts less or more severe than initially predicted.

TECHNIQUES FOR IMPACT PREDICTION

There is some confusion in many EIA texts and guides concerning the role of checklists and matrices in predicting impacts. While checklists and matrices are used to assist impact predictions by way of organizing data and information, they are *not* predictive techniques; rather, they are methods for presenting and organizing predictions. The choice of predictive technique depends on the nature of the impact indicator in question. For example, when predicting the *magnitude* or severity of an impact the approach may be qualitative and based on expert judgement. When predicting the concentration of a particular pollutant, more quantitative-based modelling techniques might be used. Techniques used for impact prediction cover a wide spectrum of economic, social, ecological, and physical factors. Morris and Therivel (2001) provide a detailed treatment of predictive techniques. While there is no single best approach to classifying predictive techniques used in EIA, there are essentially six broad classifications:

- mechanistic models;
- statistical models;
- balance models;
- experimental techniques;
- analogue techniques;
- judgemental techniques (see Sadar, 1996; Dale and English, 1999; Glasson et al., 1999; Harrop and Nixon, 1999).

These are not all of the methods and techniques available for EIA but represent a cross-section of those most commonly used.

Mechanistic Models

Mechanistic models describe cause-effect relationships in the project environment using flow diagrams or mathematical equations. Examples include Gaussian dispersion models for predicting rates and extents of pollution fallout, traffic noise generation models, population growth models, and economic impact models. Quantitative models can be divided into two categories—deterministic and stochastic.

Deterministic models depend on a fixed relationship between environmental components. For example, the **gravity model** of spatial interaction, used to predict population flows, depends on a fixed and inverse relationship between mass (population size) and distance. The system under consideration is thus at any time entirely defined by the initial conditions and model inputs chosen. With a given starting point, therefore, the outcome of the model's response is necessarily the same and is determined by the mathematical relationships incorporated in the model.

Stochastic models are probabilistic, that is, they give an indication of the *probability* of an event or events within specified spatial and temporal scales and take into consideration the presence of randomness in one or more variables. Thus, rather than giving a single point estimate, stochastic models provide a probability distribution of possible estimates or outcomes—such as the likelihood of a flood due to river control systems.

Statistical Models

The primary use of **statistical models** is to test relationships between variables and to extrapolate data. For example, based on hypothesized or proven relationships between neighbourhood housing density and quality of life, statistical models can be used to extrapolate changes in the quality of life resulting from a proposed large-scale residential housing development program. Statistical models are also used to determine whether a statistically significant difference exists between the predicted changes in impact indicators due to project influence as opposed to natural changes that would occur in the absence of project development. A word of caution is in order, however, because a measure of statistical significance, such as a 95 per cent confidence interval, does not necessarily translate directly to human perception or acceptance. For example, depending on the nature of the predicted impact, the possibility of a 5 per cent error may be deemed significant enough not to allow the action to proceed—particularly if human health and safety are at risk. A particular class of statistical models of significant value to the modelling of dynamic environments is the multivariate model, which considers changes and interactions between multiple variables within a given environmental system.

Balance Models

Mass **balance models** or mass balance equations identify inputs and outputs for specified environmental components, such as soil moisture volume or surface runoff. Balance models are essentially comprised of inputs, storage, and outputs for a particular environmental system (Box 7.1). For example, inputs to a particular system might include water and energy, whereas outputs include waste water and outflow.

Box 7.1   Conceptualization of a Simple Mass Balance Model

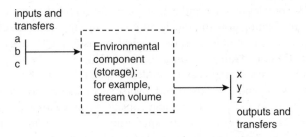

When the system is 'balanced' a + b + c = x + y + z. Changes in the environmental component can be predicted by examining potential project impacts on inputs and transfers, which will be evidenced by changes in subsequent outputs and transfers.

Balance models, such as surface hydrological models, are particularly useful for predicting physical changes in environmental phenomena where predicted changes equal the sum of the total inputs minus the sum of the total outputs (Glasson et al., 1999). A well-known example is a site water budget, which predicts the impacts of a proposed development on the input, storage, and output of water available in a particular sub-basin.

Experimental Techniques
Experimental techniques allow data or experimentation in a field or lab setting to be applied to the project environment, for example, testing the impacts of a chemical agent on a water body or predicting emissions patterns based on laboratory-controlled wind tunnels. While useful, it is important to note that not all experimental lab and field conditions transfer well to the actual project environment, particularly with regard to human conditions and impacts.

Analogue Techniques
Analogue techniques make impact predictions of a proposed development based on comparisons with the impacts of existing developments where conditions might be similar. Data might be collected and comparisons made based on literature reviews, site visits, or expert opinion. A main drawback of analogous techniques, similar to ad hoc approaches, is that in many cases experiences are not directly transferable from one project environment or one socio-economic setting to the next (Box 7.2).

Judgemental Techniques
In practice, judgemental techniques are perhaps the most commonly used predictive techniques in EIA. Judgemental techniques can be classified as expert-based or non-expert-based, examples of which include the **Delphi technique** and **intention surveys**.

Box 7.2 Analogue Techniques and the Hibernia Offshore Oil Platform Construction Project

In 1979 the Hibernia offshore oil field was discovered on the Grand Banks of Newfoundland. A proposal submitted to develop the oil field was subject to a panel review EIA under the then Canadian Federal Environmental Assessment Review process. Approval for the project was granted in 1986 and construction of a fixed gravity-base rig structure began at a dry dock in Bull Arm, Newfoundland, in 1990. The oil field is the fifth largest discovered in Canada, producing 7.6 million barrels of oil in 2004. Several companies have a share in the Hibernia megaproject: ExxonMobil Canada (33.125 per cent), Chevron Canada Resources (26.875 per cent), Petro-Canada (20 per cent), Canada Hibernia Holding Corporation (8.5 per cent), Murphy Oil (6.5 per cent), and Norsk Hydro (5 per cent).

Standing at half the height of New York's Empire State Building, the Hibernia oil-drilling platform was the first of its kind to be developed in Newfoundland, and at the time was considered one of the most significant construction projects in North America. In the absence of direct experience with such a project in Newfoundland or on Canada's east coast, the main sources of information for EIA forecasting regarding the impact of the Hibernia project on population growth, employment, and infrastructure needs in St John's, Newfoundland, were the experiences in other locales that experienced sudden growth resulting from oil exploration, namely in Aberdeen, Scotland, and Calgary, Alberta. Storey (1995) reports that comparisons of St John's with Aberdeen and Calgary led to the assumption that St John's would experience rapid population growth as a result of the Hibernia project. While such projections may have been correct for Aberdeen and Calgary, they failed to account for variations in the baseline social and institutional environment. For example, immediately following the first major oil discovery in Aberdeen in 1969, the population increased by 2.5 per cent during the first half of the 1970s. For Aberdeen, this increase was considered rapid and problematic in comparison to previous growth rates, and local institutional and infrastructural capacities were not equipped to deal with such growth. However, in St John's during that same time period the population increased by over 8 per cent, independent of oil development, and there were no major strains on institutions and infrastructure. In Calgary, immediately following oil discovery there was also a major population growth period; however, this was primarily due to local labour shortages and the labour-intensive nature of the onshore oil industry. In contrast, St John's had high unemployment rates and was engaging in relatively capital-intensive offshore oil operations. Predictions on population growth and housing demand were thus overestimated.

When making predictions based on similar projects, careful consideration must be given not only to the projects themselves but also to the nature of the biophysical and human environment in which such predictions are made.

Sources: Storey (1995); Storey and Jones (2003).

The term 'Delphi technique' was coined by Kaplan, a philosopher working with the Rand Corporation who, in 1948, headed a research effort directed at improving the use of expert predictions in policy-making (Woudenberg, 1991). The name itself refers to the ancient Greek oracle at Delphi, where those who sought advice were offered visions of the future. The Delphi technique, developed in its contemporary form during the 1950s and early 1960s, is an iterative survey-type questionnaire that solicits the advice of a group of experts, provides feedback to all participants on the statistical summaries of the responses, and provides an opportunity for each expert to revise his or her judgements (Figure 7.2). The Delphi technique is designed for use in situations where the problem does not lend itself to precise analytical techniques or where large data requirements and the lack of available quantitative data prohibit the application of traditional analytical approaches, but that can benefit from the

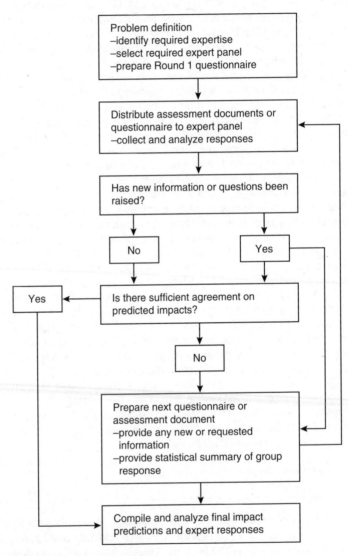

Figure 7.2 The Delphi Process

Sources: Based on Richey et al. (1985); Riggs (1983).

subjective judgements of experts on a collective basis (Dalkey, 1969). Ironically, the fact that the Delphi technique relies on expert judgement is also its primary disadvantage.

Intention surveys are often used to survey individuals about what actions they plan to take in a given situation or what they might expect to do given particular circumstances. While this approach attempts to survey and incorporate the decisions of as many people as possible, it is a costly and time-consuming exercise. In addition, citizens may have difficulty in conceptualizing how a change might affect their actions or decisions.

PREDICTING IMPACTS ON THE BIOPHYSICAL ENVIRONMENT

Predicting the impacts of project developments on the biophysical environment have been at the heart of EIA since its formal inception under the US NEPA. In fact, as discussed in Chapter 1, most EIAs during the 1970s and well into the 1980s focused almost exclusively on predicting the biophysical impacts of project development within the local project area. In current practice, biophysical impact prediction continues to play the dominant role in EIA, but its focus has broadened considerably beyond the direct and immediate project environment to include secondary and ecosystem-based biophysical change.

What to Predict
Project developments often trigger four broad areas of change that affect biophysical systems:

- biological change;
- chemical change;
- physical change;
- ecosystem change.

Based on these changes, a number of specific impacts often emerge (Table 7.1). There is no single, comprehensive list of biophysical impacts that should be considered in all EIAs, as the specific impacts to be predicted depend largely on the nature of the project and sensitivity of the local environmental setting, and are defined by the scoping process.

Techniques
It is not the purpose of this manual to provide a comprehensive discussion of the full range of techniques available for predicting biophysical environmental impacts. Numerous techniques exist for predicting such impacts, including plume models, water site budgets, wildlife population modelling, simulation modelling, dose-response factors, hydrographic extrapolation, and expert judgement. Boxes 7.3 and 7.4 provide examples of two different techniques for predicting biophysical environmental impacts. Techniques for predicting biophysical environmental impacts are given much more comprehensive treatment by Morris and Therivel (2001).

Table 7.1 Selected Biophysical Environmental Impacts: Examples of What to Predict

Air quality impacts
- pollutant concentration (various)
- pollutant dispersion
- emission levels
- emission types
- temperature
- changes in wind speed

Soil quality impacts
- erosion
- moisture
- fertility (nutrient change)
- changes in organic matter
- electrical conductivity
- chemical change
- stability
- soil pollution

Terrestrial impacts
- fragmentation
- wildlife populations
- vegetation cover and composition
- air- and water-borne pollutants
- light pollution
- disturbance
- vegetative trampling

Water impacts
- surface quantity
- surface water withdrawal
- groundwater quantity
- groundwater withdrawal
- chemical change
- turbidity
- streamflow change
- bank erosion (flood risk)
- biological change (eutrophication, algal blooms)
- pollution discharge rates (assimilative capacity)
- biological resources (fish and fish habitat)

Coastal zone impacts
- water temperature
- flooding
- alterations to tidal activity
- sedimentation
- marine resource populations
- bank or cliff stability

Note that these impacts are based on scoping issues identified in Table 6.3 to illustrate the link between scoping baseline studies and focusing impact prediction.

Box 7.3 Site Water Budgets: An Example of Predicting Sub-basin Impacts on Water Quantity

Large-scale development projects, such as thermoelectric facilities and large irrigation systems, associated with water withdrawal or changes to the hydraulic landscape have the potential to affect surface or groundwater hydrology significantly. Thus, of central concern for such projects is the storage of water in environmental systems and the flow of water between systems. While the global circulation of water is a closed system and can be characterized by the hydrological cycle, site or regional project environments are much more open, with certain inputs and outputs that control the amount of locally stored water.

The water budget can be expressed as the change in storage $\Delta S = I - O$, where $I =$ inputs and $O =$ outputs, and if $I > O$ there is an increase in water storage. Studies concerned with predicting the impacts of development projects on water storage typically focus on a particular sub-basin or geographic area from which all potential surface runoff flows

Continued

through a series of streams and rivers to particular catchments or lakes. Within the water catchment several inputs and outputs can be used to predict how changes in inputs, potentially caused by project developments and water withdrawals, might be reflected in changes in outputs. To do this, the change in storage (ΔS) is defined as (Pn + Rs + Rg) – (Et + Qs + Qg), where inputs are Pn = precipitation; Rs = surface run on; Rg = groundwater seepage; and outputs are Et = evapotranspiration; Qs = surface runoff; Qg = groundwater leakage. When an additional water usage is added to the equation, such as water withdrawal for thermoelectricity production, the site water budget becomes 'imbalanced'. Therefore, interruptions in the balance of the system due to project impacts can be predicted provided there is sufficient understanding of the local site hydrological regime.

Box 7.4 Dispersion Models: An Example of Predicting Plume Rise and Dispersion

Air pollution from industrial emissions can affect both human and physical environmental health. The dispersion of such pollutants is dependent on meteorological conditions (wind speed and patterns, insulation, cloud cover, precipitation patterns) and specific parameters of the pollution stack emissions (temperature, gas molecular structure, gas velocity). Pollutants exit a stack in the form of a 'plume', which resembles a moving cloud. The particular shape of the plume depends on local atmospheric conditions, which in turn determine the amount of dispersion. Predicting plume dispersion is critical to determining the ground-level concentration of the pollutants at various distances from the pollution source. Such information will allow project and environmental planners to determine appropriate stack height, severity of ground-level air pollution at various locations, the need for various mitigation measures, and compliance with emissions standards. Thus, of particular importance is predicting what is commonly referred to as the 'worst-case scenario'. Several techniques are available to predict plume dispersion and concentration; the most commonly applied is the Gaussian dispersion model. This model assumes that emissions spread outward from the source in an expanding plume according to the prevailing wind direction, and that the distribution of pollution concentration decreases with increasing distance from the plume. The Gaussian model is based on a specific mathematical equation that suggests that plume dispersion, pollutant concentration, and pollutant concentrations at the surface are a function of wind speed, wind direction, and atmospheric stability. One particular example of plume analysis for predicting plume height is based on Holland's equation:

$$H = (vd) / u \; (\; 1.5 + 2.68 \times 10^{-3} \, p \, ((Ts - Ta) / Ts) \, d)$$

where H = plume rise; v = stack gas exit velocity; d = stack diameter; u = wind speed; p = atmospheric pressure; Ts = stack gas temperature; and Ta = air temperature. Thus, plume height is determined by H = emission stack height.

Many types of Gaussian models are used to predict pollutant dispersion and ground-level concentration, including the Industrial Source Complex model, developed by the US Environmental Protection Agency, and the Atmospheric Dispersion Modelling System, developed by Cambridge Environmental Research Consultants UK.

Sources: Based on Elsom (2001); Harrop and Nixon (1999).

PREDICTING IMPACTS ON THE HUMAN ENVIRONMENT

Predicting the impacts of project development on the human environment is complex, uncertain, and rarely done well in EIA. While attention is often given to impacts over which the proponent has direct control, such as employment or infrastructure development, social impacts, such as quality of life or perceptions of well-being, remain the 'orphan' of the EIA process (Burdge, 2002). The role of human impact assessment in EIA is still widely debated; however, as noted at the outset of this book, 'environment' is defined here to include both the biophysical and human environments and the relations and interdependencies between them. Thus, whether human impacts are assessed separately from principal project EIAs or as part of the project EIA process, their consideration is critical to ensuring sustainability through development.

What to Predict
Project developments often trigger six broad areas of change that affect human systems:

- demographic change;
- cultural change;
- economic change;
- health and social change;
- biophysical change;
- institutional change.

Based on these changes, a number of impacts often emerge that are important to consider in project assessment (Table 7.2). That being said, there is no single, comprehensive list of human impacts that should be considered in all EIAs, as the specific impacts to be predicted depend largely on the nature of the project and the local environmental setting, and are defined by the scoping process.

Techniques
As above, it is not the purpose here to provide a comprehensive discussion of the full range of techniques available for predicting human environmental impacts. A long list of techniques exists for predicting human environmental impacts, including trends extrapolation, population multipliers, intention surveys, gravity models of spatial interaction, gaming, decision trees, traffic modelling, visibility mapping, focus group meetings, and expert judgement. Boxes 7.5 and 7.6 provide examples of two different techniques for predicting human environmental impacts. Again, techniques for predicting human environmental impacts are given much more comprehensive treatment by Morris and Therivel (2001).

CHALLENGES TO IMPACT PREDICTION

In an international study of the effectiveness of environmental assessment reports, Sadler (1996) found that 60 per cent of the time assessments are unsuccessful to only marginally successful in making precise, verifiable impact predictions and that 75 per

Table 7.2 Selected Human Environmental Impacts: Examples of What to Predict

Direct economic impacts	*Housing impacts*
• local and non-local employment	• housing demand
• labour supply and training	• public and private housing
• wage levels	• house prices
• employment demand by skill group	• homelessness and housing problems
	• density and crowding
Indirect economic impacts	
• retail expenditures	*Local service impacts*
• material and service suppliers	• educational services
• labour market pressures	• health services
• regional multiplier effects	• community services (police, fire)
• tourism	• transportation services and infrastructure
	• financial services
Demographic impacts	
• changes in population size	*Socio-cultural impacts*
• changes in population characteristics (family size, income, ethnicity)	• lifestyle changes (family life, seasonality of employment)
• changes in settlement patterns	• threats to culture and belief systems
	• perceived and actual risks
Health impacts	• social problems (crime rates, substance abuse, divorce
• quality of life (actual and perceived)	
• medical standards	• community stress (conflict, integration, cohesion)
• worker safety (risk of death or injury)	
• disease introduction and transmission	• traditional foodstuffs
• physiological impacts (stress, worker satisfaction)	• local pride and community perception
	• aesthetic impacts
• mental and physical well-being	• gender relations

Source: Based on Glasson (2001).

Box 7.5 Keynesian Multipliers: An Example of Predicting Economic Impacts

A commonly applied technique to predict the broader and often indirect economic impact of project development is the multiplier technique. A deterministic mathematical model, **multipliers** represent a quantitative expression of the extent to which some initial, exogenous change is expected to generate additional effects through linkages within the economic system. The most basic of multipliers and perhaps the most commonly used for predicting local-level impacts is the **Keynesian multiplier**. The basic principle of the Keynesian multiplier is that an injection of money into a local economic system due to project development will lead to an increase in the local-level monies by some multiple of the initial injection. For example, **Yr = KrJ**, where 'Yr' is the change in the level of income in the project region, 'J' is the initial income injection to the local economy as a result of project employment, and 'Kr' represents the regional income multiplier.

In principle, the commitment of a proponent to spend locally by way of local purchase agreements, for example, would provide an initial increase in income to the local labour

Continued

force. In turn, this would generate extra income for local suppliers, who would spend that extra income in the local region. A multiplier of 1.8, for example, would mean that for each dollar injected directly in the local economy by the project, an extra 80 cents would be generated indirectly (Morris and Therivel, 2001).

The local project environment is not, however, isolated and thus economic leakages reduce the multiplier effect. For example, some of the additional income is lost due to taxation, individuals may choose to save their extra income rather than spend it, or individuals may choose to spend that extra income in other communities or regions outside of the local environment. Thus, a multiplier for predicting local economic impacts of an initial project economic investment might more closely resemble: $Yr = [1 / 1 - (1-s)(1-tx-u)(1-m)(1-tl)] \times J$, where '$s$' is the proportion of additional income generated that is lost from the economy due to personal savings; 'tx' is the proportion of additional income that is lost due to taxes on income; 'u' is attributed to the reduction in government support and transfers as local employment increases; 'm' is the proportion of additional income spent on goods and services from outside the region; and 'tl' is the proportion of additional income that is lost due to taxes on goods and services purchased.

Typically, Keynesian multipliers are limited in EIA to the prediction of indirect economic impacts during peak and trough periods of project construction or operation, that is, best-case and worst-case scenarios.

Box 7.6   Decibel Ratings: An Example of Predicting Cumulative Noise Impacts

Noise is often an impact generated by large-scale projects such as airports, highways, railways, or heavy industries. The intensity of sound, while often expressed as 'loudness', is actually measured in decibels. The range of audible sounds is typically from 0 dB to 140 dB. However, because sound level is logarithmic in nature, adding two sounds or doubling the intensity of a sound does not result in a doubling of the decibel rating. For example, adding two identical sounds (e.g., two low-level aircraft passing overhead at once) does not double the decibel rating but increases it by only 3 dB. When two sources of sound are added together that are at different decibel ratings (e.g., sound source 1 = 59 dB; sound source 2 = 53 dB), the graph below can be used to predict the cumulative noise rating (e.g., 59 dB – 53 dB = 6 dB difference; therefore, according to the graph, add 1 dB to the higher level sound = total 60 dB). To add three sounds each of 50 dB we add 50 dB + 50 dB = 53 dB, then 50 dB + 53 dB = 54.75; the difference between 50 dB and 53 dB is 3 dB, so according to the graph we add 1.75 dB to the higher rating.

Sound level (dB)	Example	Sound level (dB)	Example
140	pain threshold for human hearing	70	busy street corner
120	pneumatic drill	60	busy office place
110	loud car horn at one metre	30	bedroom at night
105	jet flying overhead at 250 metres	10	normal breathing
90	inside a subway car	0	threshold of hearing

Continued

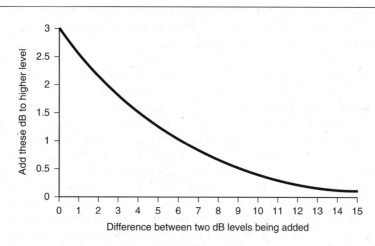

This is a rather simplistic approach to predicting noise impacts. In practice, several additional elements have to be factored into the equation, such as the level of other sounds in the environment, meteorological effects such as average wind direction and speed, local topography, other physical obstructions between the noise source and the noise receptor, and the sensitivity of the noise receptor. In addition, the impacts of noise are not limited to the human environment or to human assessments of loudness; rather, as illustrated by low-level military flight training in Labrador (Box 3.2), impacts on wildlife and the startle effect of noise are also important considerations.

Sources: Based on Harrop and Nixon (1999); Morris and Therivel (2001).

cent of the time assessments fail to indicate confidence levels of the data used for impact predictions.

In international EIA experience, little has changed since Bisset and Tomlinson (1988) noted that 697 of a total 791 predicted impacts across a survey of four UK impact assessments could not be confirmed, due in part to the vagueness of impact predictions. Similarly, Bailey et al. (1992), in an environmental audit of waterway developments in Western Australia, found that 90 per cent of impact predictions did not indicate a specific time scale in which the predicted impact was expected to occur. This is confirmed by Morrison-Saunders and Bailey (1999), who more recently analyzed six Australian cases and found little evidence of impact quantification or precision in prediction, with most predictions being vague and qualitative in nature. In the Canadian experience, Beanlands and Duinker (1983) examined the scientific quality of EIA in Canada from an ecological perspective. Their analysis suggested the limited value of impact predictions for decision-making and subsequent testing and replication. To address this issue, EIA practitioners need to recognize that uncertainty is inevitable and that impact predictions must be stated in such a way that they can be verified.

Addressing Uncertainty in Impact Prediction

The question of any impact prediction is 'how certain are we that the predicted impact is correct or will actually occur?' Uncertainty is inevitable in impact predictions, since no model or predictive technique can accurately predict what might happen in a dynamic environment. There are, however, at least three principal measures that can be taken to address uncertainty in impact prediction:

1. **Probability analysis** involves quantifying the probability that an impact or range of impacts is likely to occur and under what conditions.
2. **Sensitivity analysis** examines the sensitivity of the prediction to minor variations in input data, environmental parameters, and assumptions. For example, one might examine how sensitive a prediction concerning the impacts of worker influx might be on a local population during the construction phase of a project by testing a variety of different scenarios concerning labour demands and the composition of the local workforce. Sophisticated techniques are available for evaluating the sensitivity of impact predictions, such as **Monte Carlo analysis**, a mathematical technique using random samples, while others simply involve providing a range of impact predictions for a particular impact indicator based on minor variations in the data and parameters on which those predictions are based.
3. **Confirmatory analysis** is used to test for uncertainty concerning the predictive technique itself and to ensure that the predicted impact is not solely a product of the particular technique used. In other words, the objective is to determine the extent to which different techniques provide similar predicted outcomes.

Verifiable Impact Predictions

A critical characteristic to the value of impact predictions is that they be stated in such a way that they can actually be followed up and verified. This can be accomplished by one of two methods, depending on the nature and availability of baseline data.

Hypotheses-based approaches. When impacts are predicted in environmental assessment, they should, where at all possible, be based on rigorous and falsifiable null-impact hypotheses stating the relevant affected variables, impact magnitude, spatial and temporal extent, probability of occurrence, significance, and associated confidence intervals. Ringold et al. (1996) suggest that impact predictions should be couched in quantitative terms, identify the spatial and temporal characteristics of interest, and state the Type I and Type II probabilities. This is consistent with recommendations of the panel reviewing the recent Voisey's Bay nickel mine project proposal for northern Labrador, which suggested that a hypothesis-driven approach is the preferred formula for impact prediction (Voisey's Bay Panel, 1999). Such an approach to impact prediction allows practitioners to measure variables and either accept or reject impact predictions and environmental change within specified confidence limits. However, it is important to note that a hypothesis-driven approach to impact prediction does not relieve the role of professional judgement when considering what constitutes a significant or acceptable deviation from the null, and it inevitably relies on the availability and quality of data.

A hypothesis-driven approach requires that assumptions underpinning impact predictions be clearly stated, including any exogenous factors associated with the impact prediction. Munro et al. (1986), for example, identify a range of probability and confidence intervals associated with impact predictions (Table 7.3). The lower the associated confidence level, the greater is the likelihood that effects monitoring will become important as the project proceeds.

Threshold-based predictions. Impact prediction typically involves the use of some model either mathematically or conceptually representing the biophysical or socio-economic environment. Uncertainties emerge, contributing to inaccurate impact predictions because the ultimate purpose of such models is to simplify the actual system and, inherently, they cannot capture the multiplicity of factors interacting and affecting the natural system (De Jongh, 1988). In any environmental or socio-economic system, several different processes are often involved that may affect the variable or environmental component of concern. Thus, impact predictions often turn out to be inaccurate because of the mix of assumptions that normally have to be made and the multiplicity of exogenous factors involved (Mitchell, 2002: 56). First, projects are almost always modified during the development phase, thereby making initial impact predictions less valuable from a follow-up and learning perspective. Second, given the nature of constantly changing, and often unpredictable, human environmental systems, the environmental impacts of development can rarely be predicted with any degree of certainty. This is particularly true in the social environment, where humans often react and adapt in anticipation of project-induced change.

However, not all environmental effects need to be predicted precisely; in many cases, particularly where there are inadequate baseline data, it is recognized in the assessment that outcomes should not exceed specified threshold levels, in which case management practices, the setting of targets, and the determination of threshold levels become the focus of attention (Storey and Noble, 2004). It is often the case, for example, that a threshold or particular target is set and a monitoring framework is established to ensure that negative impacts do not exceed certain thresholds and that positive impacts meet specified expectations. Where hypothesis-based approaches are not suitable, **threshold-based predictions** may rely on previous experiences with similar projects, similar types of impacts in different environments, or regulatory standards. For example, it is known that certain levels of noise can have a negative effect on human hearing and certain project construction activities (such as heavy machinery or blasting) generate particular levels of noise. One approach would be to

Table 7.3 Impact Predictions and Confidence Limits

Confidence levels	Data characteristics	Knowledge	Permitted approach
high/factual (95%)	reliable	proven cause-effect relationship	statistical prediction
fairly high	sufficient	evidence for hypotheses	quantitative simulation
fairly low	insufficient	postulated linkages	conceptual modelling
low/intuitive (50%)	absent or unreliable	speculation	professional opinion

Source: Munro et al. (1986).

compare the level of cumulative noise to specified thresholds for human hearing in order to predict the impact of project construction activities.

Where comparative examples or regulatory standards are not readily available, an alternative approach is to base impact predictions on a desirable level or maximum levels of change. **Maximum allowable effects levels**, for example, reflect an approach to impact prediction where the impact is stated in the form of a hypothesis such as 'project impact *i* will not exceed a particular threshold or desired effects level for a particular impact indicator *j*.' This approach was used in the Hibernia offshore oil project for predicting potential project impacts on the biophysical environment, and it is particularly useful for managing project outcomes and meeting desired sustainability objectives.

KEY TERMS

accuracy
balance model
confirmatory analysis
Delphi technique
deterministic model
gravity model
intention survey
Keynesian multiplier
mechanistic model

maximum allowable effects level
Monte Carlo analysis
multiplier
precision
probability analysis
sensitivity analysis
statistical model
stochastic model
threshold based prediction

STUDY QUESTIONS AND EXERCISES

1. Discuss the relationship between accuracy and precision in impact prediction.
2. What are some of the challenges to making impact predictions concerning biophysical and social change?
3. What are the advantages and disadvantages to analogue techniques for impact prediction?
4. Expert judgement is perhaps the most common predictive element in most EIAs. Discuss the advantages and limitations of relying on expert judgement for impact prediction.
5. Consider a condition with the following baseline noise environment:

   - average daily road traffic　　55 dB
   - average daily rail traffic　　　48 dB
   - average daily air traffic　　　 58 dB
   - average noise from factories 62 dB

   What would be the total noise impact of a proposal to develop a heavy manufacturing plant in the region with equipment operating at average daily noise levels of 68 dB? Would you consider this to be a significant impact? Explain. How might you manage such an impact?
6. Obtain a completed project EIS from your local library or government registry, or access one on-line. Scan the document for stated impact predictions and examine how these predictions are stated. Are the statements based on verifiable hypotheses? Are thresholds or maximum allowable effects levels stated? Given the nature of the predictions and the way in which they are stated, do you think that they can be followed up and verified? Compare your findings to those of others.

7. Obtain a completed project EIS from your local library or government registry, or access one on-line. Scan the document and generate a list of the techniques used to predict, describe, or assess impacts on the biophysical and human environments. Compare your results with those of others. Is there a common set of techniques that emerge for various environmental components, such as water quality, air quality, or employment?

REFERENCES

Bailey, J., V. Hobbs, and A. Saunders. 1992. 'Environmental Auditing: Artificial Waterway Developments in Western Australia', *Journal of Environmental Management* 34: 1–13.

Beanlands, G.E., and P.N. Duinker. 1983. 'Lessons from a Decade of Offshore Environmental Impact Assessment', *Ocean Management* 9, 3/4: 157–75.

Bisset, R., and P. Tomlinson. 1988. 'Monitoring and Auditing of Impacts', in P. Wathern, ed., *Environmental Impact Assessment: Theory and Practice*. London: Unwin Hyman, 117–26.

Buckley, R. 1991. 'Auditing the Precision and Accuracy of Environmental Impact Predictions in Australia', *Environmental Monitoring and Assessment* 18: 1–23.

Burdge, R. 2002. 'Why Is Social Impact Assessment the Orphan of the Assessment Process?', *Impact Assessment and Project Appraisal* 20, 1: 3–11.

Dale, V., and M.R. English, eds. 1999. *Tools to Aid Environmental Decision-Making*. New York: Springer.

Dalkey, N. 1969. 'An Experimental Study of Group Opinion: The Delphi Method', *Futures* 1: 408–20.

De Jongh, P. 1988. 'Uncertainty in EIA', in P. Wathern, ed., *Environmental Impact Assessment: Theory and Practice*. London: Unwin Hyman.

Duinker, P.N., and G.E. Baskerville. 1986. 'The Significance of Environmental Impacts: An Exploration of the Concept', *Environmental Management* 10, 1: 1–10.

Elsom, D.E. 2001. 'Air Quality and Climate', in Morris and Therivel (2001).

Glasson, J. 2001. 'Socioeconomic Impacts 1: Overview and Economic Impacts', in Morris and Therivel (2001).

———, R. Therivel, and A. Chadwick. 1999. *Introduction to Environmental Impact Assessment: Principles and Procedures, Process, Practice and Prospects*, 2nd edn. London: University College London Press.

Harrop, D.O., and A.J. Nixon. 1999. *Environmental Assessment in Practice*. Routledge Environmental Management Series. London: Routledge.

Mitchell, B. 2002. *Resource and Environmental Management*, 2nd edn. New York: Prentice-Hall.

Morris, P., and R. Therivel, eds. 2001. *Methods of Environmental Impact Assessment*, 2nd edn. London: Taylor and Francis Group.

Morrison-Saunders, A., and J. Bailey. 1999. 'Exploring the EIA/Environmental Management Relationship', *Environmental Management* 24, 3: 281–95.

Munro, D., T. Bryant, and A. Matte-Baker. 1986. *Learning from Experience: A State of the Art Review of Environmental Impact Assessment Audits*. Ottawa: Canadian Environmental Assessment Research Council.

Richey, J.S., B.W. Mar, and R. Horner. 1985. 'The Delphi Technique in Environmental Assessment: Implementation and Effectiveness', *Journal of Environmental Management* 1: 135–46.

Riggs, W.E. 1983. 'The Delphi Technique: An Experimental Evaluation', *Technological Forecasting and Social Change* 23: 89–94.

Ringold, P.R., J. Alegria, R.L. Czaplewski, B.S. Mulder, T. Tolle, and K. Burnett. 1996. 'Adaptive Monitoring Design for Ecosystem Management', *Ecological Applications* 6, 3: 745–7.

Sadar, H. 1996. *Environmental Impact Assessment*, 2nd edn. Ottawa: Carleton University Press.

Sadler, B. 1996. *Environmental Assessment in a Changing World: Evaluating Practice to Improve Performance*. Final report of the International Study of the Effectiveness of Environmental Assessment. Fargo, ND: IAIA.

Storey, K. 1995. 'Managing the Impacts of Hibernia: A Mid-Term Report', in B. Mitchell, ed., *Resource and Environmental Management in Canada*, 2nd edn. Toronto: Oxford University Press, 310–34.

———— and P. Jones. 2003. 'Social Impact Assessment, Impact Management and Follow-up: A Case Study of the Construction of the Hibernia Offshore Platform', *Impact Assessment and Project Appraisal* 21, 2: 99–107.

———— and B. Noble. 2004. *Toward Increasing the Utility of Follow-up in Canadian EA: A Review of Concepts, Requirements and Experience*. Report prepared for the Canadian Environmental Assessment Agency. Hull, Que.: CEAA.

Voisey's Bay Mine and Mill Environmental Assessment Panel. 1999. *Report on the Proposed Voisey's Bay Mine and Mill Project / Environmental Assessment Panel*. Hull, Que.: Canadian Environmental Assessment Agency.

Woudenberg, F. 1991. 'An Evaluation of Delphi', *Technological Forecasting and Social Change* 40: 131–50.

Determining Impact Significance

IMPACT SIGNIFICANCE

An important, yet highly subjective part of EIA is determination of impact significance. Section 4(c) of the Canadian Environmental Assessment Act states that one of
the purposes of the Act is to ensure that projects do not cause significant adverse environmental effects outside the jurisdictions in which the projects are carried out.
Determination of impact significance essentially involves making judgements about
the importance of impacts. Significance reflects the degree of importance placed on
the impact in question and consists of two components:

- the significance of predicted project impacts;
- the significance of predicted impacts following impact management or mitigative measures.

Impacts associated with the latter are typically referred to as **residual impacts**, or
impacts that remain after all management and mitigation measures have been implemented.

The determination of significance begins at the outset of the EIA process, when a
decision is being made as to whether the proposal requires a formal assessment, and
extends throughout the scoping, prediction, mitigation, and follow-up stages.
Lawrence (2004) suggests that interpretations of significance are involved when
deciding or determining:

- whether EIA requirements are to be applied;
- which EIA requirements are to be applied;
- the focus of the EIA;
- which aspects are important;
- which activities are most likely to generate adverse environmental effects;
- which options to the proposed action are available;
- which impacts to focus on in the course of further analysis;
- which impacts are most important;
- whether management measures are necessary;
- which management measures should be employed;
- which project cumulative effects are important;
- the roles of various agencies and the public;

- whether proposed project actions or activities are acceptable;
- which conditions to impose;
- whether to monitor project impacts;
- which impacts to actually monitor.

DETERMINING SIGNIFICANCE

Notwithstanding the importance of determining impact significance, there is little consistent guidance for practitioners. Generally speaking, impacts are likely to be considered significant if they are:

- adverse;
- intensive in concentration or associated with significant levels of change;
- associated with a high degree of probability;
- frequent and long-lasting;
- likely to occur at a broad spatial scale;
- irreversible;
- associated with cumulative change;
- going to detract from the sustainability of environmental and socio-economic systems;
- likely to affect ecological functions or exceed **assimilative capacity** of the environment;
- associated with variables of societal importance and public concern;
- not in compliance with existing standards or regulations;
- likely to exceed desired levels of change.

Impact significance, then, is a function of the characteristics of the impact and the impact importance or value attached to the affected component (Figure 8.1).

Impact Characteristics

For each impact prediction, several questions concerning the characteristics of impacts are typically asked (see Lawrence, 2004; Canter, 1996; Wood, 1995). The specific nature of these questions varies from project to project but usually involves the identification of 'major', 'moderate', 'minor', and 'negligible' effects based on several impact characteristics (Table 8.1). Many of these characteristics were discussed in Chapter 3 and are reviewed only briefly here, with particular attention to other impact characteristics not previously introduced.

Adverse impact. A first step in determining impact significance is to determine whether impacts are likely to be adverse following impact mitigation. Some major

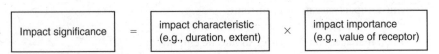

Figure 8.1 Definition of Impact Significance

Table 8.1 Selected Impact Significance Criteria for the Ekati Diamond Mine EIA

Impact significance	Type of environmental component		
	Physical	Biological	Socio-economic
Major	Parameter affected within most of ecozone for several decades	Whole stock or population of ecozone affected over several generations	Whole population of region affected over several generations
Moderate	Parameter affected within most of ecoregion for one or more decades	Portion of population of ecoregion affected over one or more generations	Community affected over one or more generations
Minor	Parameter affected within most of ecoregion during less than one decade	A specific group of individuals within an ecosystem affected during less than one generation	A specific group of individuals within a community affected during less than one generation
Negligible	Parameter affected within some part of ecoregion for a short time period	A specific group of individuals within an ecosregion affected for a short time period	A specific group of individuals within a community affected for a short time period

Source: Based on BHPB (1998).

factors that should be used to determine whether environmental effects are adverse are described by FEARO (1994) and summarized in Table 8.2. The most common way of determining whether effects are adverse is to compare the quality of the environment with and without the project, using as variables the relevant environmental components identified in Table 8.2. This does, however, imply that baseline data are available or can be established for each of the variables concerned.

Magnitude. The degree or amount of change associated with an adverse environmental impact, magnitude is usually not considered to be high if a major adverse impact can be mitigated. It is important to emphasize that magnitude itself does not always equate directly with significance. For example, a large increase in mercury levels in a local water supply reservoir may not be considered significant if the mercury levels are still within the quality standards for drinking water. The significance of an impact, in terms of its magnitude, is typically assessed in comparison to deviation from a pre-project baseline condition or from an alternative predetermined measurement point. For example, in the Jack Pine mine EIS, impact magnitude measures for noise, groundwater hydrology, surface water hydrology, and wildlife health were determined based on specified measured baseline conditions (Table 8.3).

Probability. If an impact is not likely to happen, then it may not be significant. The determination of likelihood is based on two criteria: (1) the probability of occurrence and (2) scientific certainty. In practice, the likelihood of an attribute of significance

Table 8.2 Selected Factors in Determining Adverse Environmental Impacts

Changes in the environment	Human impacts resulting from changes
Negative effects on the health of biota, including plants, animals, and fish	Negative effects on human health, well-being, or quality of life
Threat to rare or endangered species	Increase in unemployment or shrinkage in the economy
Reduction in species diversity or disruption of food webs	Reduction of the quality or quantity of recreational opportunities or amenities
Loss or damage to habitats, including habitat fragmentation	Detrimental change in the current use of lands and resources for traditional purposes by Aboriginal persons
Discharges or release of persistent and/or toxic chemicals, microbiological agents, nutrients (e.g., nitrogen, phosphorus), radiation, or thermal energy (e.g., cooling waste water)	Negative effects on historical, archaeological, paleontological, or architectural resources
Population declines, particularly in top predator, large, or long-lived species	Decreased aesthetic appeal or changes in visual amenities (e.g., views)
The removal of resource materials (e.g., peat coal) from the environment	Loss or damage to commercial species or resources
Transformation of natural landscapes	Foreclosure of future resource use or production
Obstruction of migration or passage of wildlife	
Negative effects on the quality and/or quantity of the biophysical environment (e.g., surface water, groundwater, soil, land, and air)	

Source: FEARO (1994).

is often rated on a scale of 'none' (0 per cent chance), 'low' (< 25 per cent chance), 'moderate' (25–75 per cent chance), and 'high' (> 75 per cent chance). While sometimes based on **statistical significance**, this is only one means of determining probability (Table 8.4).

Duration and frequency of impact. Significance may be based on duration of the impact—short-term (1–5 years after project completion), medium-term (6–15 years after completion), or long-term (more than 15 years). This is, however, a highly subjective process and should be based on similar activities and impacts from comparable projects where possible. It is also important to consider whether the impact is continuous, delayed, or immediate. An impact likely to be delayed for 10 years after project completion may or may not be considered as significant as an immediate and continuous impact.

Spatial extent of impact. As depicted in Table 8.1, localized impacts, such as noise, are often not considered as significant as impacts experienced at locations far removed from the project, such as **acid mine drainage** or atmospheric emissions.

Table 8.3   Selected Impact Magnitude Criteria for the Jack Pine Mine EIA

Factor	Impact magnitude criteria and indices
Noise	Negligible (0): no projected increase in ambient noise levels Low (+5): increased noise levels do exist but do not exceed the nighttime noise criteria Moderate (+10): increased noise levels exceed to the nighttime criteria High (+15): increased noise levels exceed the daytime noise criteria
Groundwater hydrology	Negligible (0): no change from the baseline case Low (+5): near (slightly above) the baseline case Moderate (+10): above the baseline case High (+15): substantially above the baseline case
Surface water hydrology	Negligible (0): < 5 per cent change Low (+5): 5–10 per cent change Moderate (+10): 10–30 per cent change High (+15): > 30 per cent change
Wildlife health	Negligible (0): no appreciable increase in hazard compared to baseline Low (+5): possible small increase in hazard compared to baseline Moderate (+10): possible moderate increase in hazard compared to baseline High (+15): substantial increase in hazard compared to baseline

Source: Shell Canada Limited (2002).

Reversibility. Reversible impacts are typically considered less significant than irreversible ones.

Standards and regulations. Predicted impacts or impacts following mitigation are often compared against environmental standards or regulations—in essence, specified thresholds. A threshold involves a clearly defined performance level, usually applied to distinguish between significant and insignificant effects. Standards and regulations are perhaps the most common approach to thresholds for assessing impact significance. Environmental impacts within specified standards or regulations are often deemed to be insignificant in comparison to those impacts that exceed regulations or, for example, those that attain maximum allowable concentrations. Many jurisdictions have standards for drinking water quality or levels of industry emissions. Lawrence (2004) identifies at least three major types of standards or regulatory thresholds:

- *exclusionary*: leads to automatic rejection of a proposal;
- *mandatory*: leads to a mandatory finding of significance;
- *probable*: normally significant but subject to confirmation.

However, with the exceptions of California and Australia, explains Lawrence, significance thresholds are not normally included in EIA requirements.

Table 8.4 Impact Probability Classifications

High	Previous research, knowledge, or experience indicates that the environmental component *has experienced* the same impact from activities of similar types of projects.
Moderate	Previous research, knowledge, or experience indicates that the environmental component *may have experienced* the same impact from activities of similar types of projects.
Low	Previous research, knowledge, or experience indicates there is a *small likelihood* that the environmental component has experienced the same impact from activities of similar types of projects.
Unknown	There is *insufficient research, knowledge, or experience* to indicate whether the environmental component has experienced the same impact from activities of similar types of projects.

Source: Based on BHPB (1998).

Cumulative impacts. Impacts that are cumulative in nature, that is, they add to or interact with an already existing adverse impact to detract additively or synergistically from environmental quality, are more likely to be deemed significant compared to those impacts that do not affect already affected environmental components.

Sustainability criteria. EIA should contribute to the sustainable development of the environment. Thus, an alternative approach to determining impact significance is to ask whether the project's impacts make a positive contribution to or detract from sustainability. Gibson (2001), for example, identifies several generic sustainability-based questions for evaluating the significance of environmental impacts, including:

- Could the effect add to stress that might undermine ecological integrity or damage life-support functions?
- Could the effects contribute to ecological rehabilitation or otherwise reduce stress on environmental components?
- Could the effects increase equity in the provision of material security, including present and future generations?
- Could the effects provide more opportunities for economic well-being while reducing material and energy demands?

In the case of the Voisey's Bay mine project, introduced in Chapter 1, the 'sustainability of resources', defined as the capacity of an affected resource to meet present and future needs (Voisey's Bay Nickel Company, 1997), was used as a factor to assist in determining the significance of residual impacts based on such concepts as ecosystem integrity, carrying capacity, and assimilative capacity (Table 8.5).

Impact Importance
The importance of particular impacts is related to ecological and societal values, and in each of these instances desired or acceptable thresholds of impacts must be determined.

Table 8.5 Sustainability Criteria Used in the Voisey's Bay EIS for Significance Determination

Sustainability rating	Sustainable use of renewable resources criteria
High	Previous research/experience indicates that the environmental effect on the VEC would not reduce biodiversity or the capacity of resources to meet present and future needs.
Moderate	Previous research/experience indicates that the environmental effect on the VEC may, to a certain extent, reduce biodiversity or the capacity of resources to meet present and future needs.
Low	Previous research/experience indicates that the environmental effect on the VEC would reduce biodiversity or the capacity of resources to meet present and future needs.
Nil	Previous research/experience indicates that the environmental effect on the VEC would eliminate biodiversity or the capacity of resources to meet present and future needs.
Unknown	There is insufficient research/experience to indicate whether the environmental effect on the VEC would reduce biodiversity or the capacity of resources to meet present and future needs.

Source: Voisey's Bay Nickel Company (1997).

Ecological value. Criteria include aspects of the environment that are critical to ecosystem functioning, and may include:

- effects on plant and animal habitat;
- endangerment to rare and endangered species;
- ecosystem resilience, sensitivity, biodiversity, and carrying capacity;
- viability of local species populations.

In addition, it is important to consider whether the effects will occur in areas that have previously been adversely affected by human activities or are ecologically sensitive.

Societal value. Project impacts on the biophysical environment often affect factors of importance to society and human life, as well as those biophysical components that are valued for aesthetic or sentimental reasons. Societal values that can be affected include:

- human health and safety;
- potential loss of green space;
- recreational value;
- demands on public resources;
- demands for infrastructure and services;
- demographic effects.

Such impacts are likely to be considered significant.

Impacts of high public concern, such as risks to human health, are likely to be considered more significant than those not of public concern or associated with low levels of risk. Conventional levels for 'acceptable risk' to the public range from one in 10,000 to one in 10 million. In addition to actual risk, impact significance can be determined by assessing public concerns or perceived risk. Simply because risk assessment may suggest a low level of actual risk, perceived levels of risk may be high due to incomplete public knowledge and understanding or as a result of previous incidents. Typical examples of projects where perceived risk might play an important role in determining impact significance include nuclear energy generation and hazardous waste disposal projects.

Desired thresholds. Thresholds need not always be based on standards or regulatory requirements. It is often the case, as already discussed, that thresholds for adverse impacts are stated as *maximum allowable effects levels.* These are often based on the desired outcomes and are determined by policy, compatibility with existing plans or programs, or societal objectives. In cases where predicted impacts exceed desired thresholds or maximum allowable change for particular environmental components, the impact is considered significant—even though it may not exceed regulatory standards or thresholds.

METHODS, TECHNIQUES, AND GUIDANCE

Determining impact significance is a highly subjective process, yet significance can and should be determined in a consistent and systematic fashion. Several methods and techniques are used to determine and interpret impact significance: data collection and compilation procedures, such as site visits, geographic information systems, simulation modelling, and statistical significance tests; data scaling and screening procedures, such as threshold analysis and constraint mapping; qualitative and quantitative aggregation and evaluation procedures, including concordance analysis, multi-criteria analysis, ranking and weighting, and risk assessment; and formal and informal public interaction procedures, such as open houses, workshops, and advisory committees (Lawrence, 2004). Similar to many other aspects of EIA practice, there is no specific set of methods and techniques for determining significance. A number of guidelines, however, should be adhered to, namely that significance thresholds, criteria, and methods be explicit, easy to use, traceable, verifiable, readily understandable, relevant to the problem at hand, and capable of facilitating the interpretation of impact significance (Sadler, 1996).

Whether a predicted impact meets a pre-established standard or threshold makes determining impact significance relatively straightforward. However, in many cases impacts are evaluated only in terms of relative significance. For example, is a short-term impact on human health from factory emissions more significant than a major long-term increase in local employment? It is often the case in EIA that impacts are expressed as 'major' or 'minor' without evaluating the relative importance of the environmental components under consideration (Figure 8.2). Yet, without some type of weighting approach to determine the **relative significance** of all the environmental

Significance of residual effects

blank	not applicable
○	negligible
●	negligible/minor
◉	minor
◎	minor/moderate
◆	moderate
✳	moderate/major
✸	major
⊗	unknown

Project Period and Activity

Construction Phase

VECs	blasting	road traffic	roads, dams, infrastructure	quarrying	noise	channel diversion	lake dewatering	pre-stripping	diesel power generation	process plant	human activity	tailings disposal	flow and stream diversion
air quality	○	○											
permafrost			○										
eskers				◉									
water quality			●			○			○				
fish and aquatic habitat			◉				●			●			
vegetation			○					●		●			
wildlife and wildlife habitat			●			●							
caribou			●		●						●	●	
grizzly bears			◉		◉						●	●	
wilderness											◉		
biodiversity											○		
hydrology			○			○	○						◉
climate							○						
groundwater							○						

Operations Phase

VECs	process plant operation	diesel power generation	roads and road traffic	tailings water disposal	freshwater supply	waste rock dumping	excavation of pits	winter roads	human activity	diversion channel	noise	flow and stream diversion
air quality	○	○	○									
permafrost				○								
eskers												
water quality				○		○						
fish and aquatic habitat			●		●							
vegetation		○					●	◉				
wildlife and wildlife habitat												
caribou	○		○	○					●	●	●	
grizzly bears	○		●						◉		◉	
wilderness	●		◉						◉			
biodiversity									○			
hydrology						⊗						◉
climate		○										
groundwater												○

Figure 8.2 Sample Impact Significance Matrix

Source: Based on BHPB (1998).

components, comparing project action 'i' on environmental components 'x' and 'y' gives little indication of actual significance (Figures 8.3 and 8.4). Several methods are available for determining the relative importance of environmental impacts, including **fixed-point scoring**, **rating**, and **paired comparisons**. A more detailed discussion of weighting methods and techniques is provided by Hajkowicz et al. (2000).

Fixed-Point Scoring

In fixed-point scoring, a fixed number of values are distributed among all affected environmental components. The higher the point score the more important the environmental component. In other words, impacts on environmental components assigned high scores are likely to be considered more significant than impacts on environmental components with low fixed-point scores. For example, assume four VECs for which a fixed number of points, totalling no more than '1', must be distributed to indicate relative VEC importance:

VEC	Importance ($w_i / 1$)
water quality	.25
noise	.10
employment	.30
human health	.35

Increasing the importance of any one VEC directly affects the relative significance of impacts on that VEC. At the same time, those assigning the weights, whether EIA

Environmental parameter	Project Actions						Impact Score
	blasting	site clearing	dredging	road construction	waste disposal	equipment transport	
Air quality	−1			−1	−1		−3
Water resources	−2	−3	−3				−8
Water quality	−2	−4	−2				−8
Noise	−2		−1	−2		−2	−7
Forests and vegetation		−5		−3			−8
Wildife	−2	−4		−2			−8
Human health	−2			−1	−4		−7

+ = positive impact	no impact =	moderate impact = 3
− = adverse impact	negligible impact = 1	major impact (irreversible or long term) = 4
	minor (slight or short term) = 2	severe impact (permanent) = 5

Figure 8.3 Unweighted Impact Assessment Matrix. Assuming an example of simple additive impacts, the impact of 'site clearing' on 'forest and vegetation' appears to be the most significant individual project impact. In terms of overall project impacts, the impacts of project activities on water resources, water quality, forests and vegetation, and wildlife appear to be equally significant. The relative importances of the affected environmental components have not been determined.

Environmental Parameter	Weight	blasting	site clearing	dredging	road construction	waste disposal	equipment transport	Impact Score
				Project Actions				
Air quality	0.16	−1(0.16) = −0.16			−1(0.16) = −0.16	−1(0.16) = −0.16		−0.48
Water resources	0.08	−2(0.08) = −0.16	−3(0.08) = −0.24	−3(0.08) = −0.24				−0.64
Water quality	0.22	−2(0.22) = −0.44	−4(0.22) = −0.88	−2(0.22) = −0.44				−1.76
Noise	0.04	−2(0.04) = −0.08		−1(0.04) = −0.04	−2(0.04) = −0.08		−2(0.04) = −0.08	−0.28
Forests and vegetation	0.08		−5(0.08) = −0.40		−3(0.08) = −0.24			−0.64
Wildife	0.08	−2(0.08) = −0.16	−4(0.08) = −0.32		−2(0.08) = −0.16			−0.64
Human health	0.22	−2(0.22) = −0.44			−1(0.22) = −0.22	−4(0.22) = −0.88		−1.54

+ = positive impact
− = adverse impact

no impact =
negligible impact = 1
minor (slight or short term) = 2

moderate impact = 3
major impact (irreversible or long term) = 4
severe impact (permanent) = 5

Figure 8.4 Weighted Impact Assessment Matrix. Weights are assigned to indicate the importance of the affected environmental components and therefore the relative significance of the impacts. Using the same example as in Figure 8.3, the impact of 'site clearing' on 'forest and vegetation' is no longer the most significant individual project impact. Given the importance of the affected components, the impacts of 'site clearing' on 'water quality' and of 'waste disposal' on 'human health' are now the most significant individual impacts. In terms of overall project impacts, the impacts of project activities on 'water quality' are now the most significant, whereas overall impacts on 'forest resources' and 'wildlife' now appear to be relatively less significant. As a result, the unweighted assessment matrix depicted in Figure 8.3 for determining significance may lead to an incorrect allocation of resources for impact management.

decision-makers, experts, or public interest groups, are forced to make trade-offs between VECs because increasing the importance of one VEC requires decreasing the importance of another. While explicit for decision-making purposes, direct trade-offs between VECs may not always be possible.

Rating

With a rating approach to assigning impact importance, the importance or significance of each VEC is indicated on a numerical rating scale, for example: 1 = not important; 5 = moderately important; 10 = extremely important; 2, 3, 4, 6, 7, 8, 9 = intermediate values. While rating does not require direct trade-offs in assignment of weights, the disadvantage is that this does not directly indicate the relative importance of one component in comparison to another.

Paired Comparisons

Similar to fixed-point scoring, paired comparisons force the decision-maker to consider trade-offs. However, whereas fixed-point scoring requires multiple, simultaneous, and often complex trade-offs for the entire list of environmental components, paired comparisons involve making trade-offs one at a time, for each pair of VECs, thereby contributing to better overall understanding of the decision problem.

The paired comparison approach is based on Saaty's Analytical Hierarchy Process (AHP), a systematic procedure for representing the elements of any problem hierarchically. The AHP organizes the basic rationality by breaking down a problem into its smaller constituent parts and then guides the decision-maker through a series of pairwise comparison judgements to express the relative importance of the elements in hierarchy in ratio form, from which decision weights are derived based on the principal eigenvector approach (Saaty, 1977). An eigenvector is a linear combination of variables that consolidates the variance, or eigenvalues, in a matrix. The eigenvalues indicate the relative strength (weight) of each of the VECs, where the larger the eigenvalue the larger the role the paired comparison plays in weighting the entire assessment matrix and therefore determining significance.

For any pair of environmental components or VECs i and j out of the set of components C, the individual decision-maker can provide a paired comparison C_{ij} of the components under consideration on a ratio scale that is reciprocal, such that $c_{ji} = 1/c_{ij}$. In other words, if environmental component VEC i is 'seven times' more important than VEC j, then the reciprocal property must hold, that is, VEC j must be seven times less important than VEC i. When a decision-maker compares any two VECs i, j, one VEC can never be judged to be infinitely more important than another. If such a case should arise where VEC i is infinitely more than j, then no decision tool would be required.

Using the paired comparison approach, decision-makers are presented with a nine-point decision scale ranging from '1', if both components or VECs are equally preferred, '3' for a weak preference of VEC i over j, '5' for a strong preference and so forth. If VEC j is preferred to i for any given criteria, the reciprocal values hold true: 1, 1/3 and 1/5 (Box 8.1). The scale is standardized and unit free, thus there is no need to transform all measures, for example, into monetary units for comparative purposes, which is a key advantage of paired comparisons over cost-benefit analysis or utility functions. An additional advantage of the paired comparisons over other weighting methods is that the paired comparison approach also provides a measure of consistency or the extent to which weights were assigned purposefully or at random.

While paired comparison generates the most informative weightings for determination of significance, it is also the most complex of the three approaches presented here. In fact, when relying on the public or experts to assign the values, the approach is limited to the number of items that an individual can simultaneously compare— usually 7 +/- 2 (Miller, 1956). However, when relying on simulation models or computer-based technology such as geographic information systems, the number of environmental components that can be weighted is limited only by the time and resources available to the EIA practitioner.

Box 8.1   Paired Comparison Process for Assigning Weights to Environmental Components

The relative importance of component c_i in the assessment of project impacts is defined by:

9 = component i is extremely more important than j
7 = component i is very important compared to component j
5 = component i is strongly more important than component j
3 = component i is moderately more important than component j
1 = component i is equally important to component j
1/3 = component j is moderately more important than component i
1/5 = component j is strongly more important than component i
1/7 = component j is very important compared to component i
1/9 = component j is extremely more important than component i

Sample paired comparison assessment matrix for environmental components (VECs)

	VEC A	VEC B	VEC C
VEC A	1	1/3	7
VEC B	1	1	7
VEC C	1/7	1/7	1

For example, VEC 'B' is considered 'moderately more important' than VEC 'A' so a value of '3' is entered in row 2, column 1; thus 1/3 in row 1, column 2.

To derive the weights for each VEC:

1. Divide each cell entry of the matrix by the sum of its corresponding column to normalize the matrix:

	VEC A	VEC B	VEC C
VEC A	0.24	0.23	0.47
VEC B	0.72	0.68	0.47
VEC C	0.03	0.10	0.07

2. Determine the priority vector of the matrix by averaging the row entries in the normalized assessment matrix.
   Priority vector: A = 0.31; B = 0.62; C = 0.06

3. Multiply each column of the initial paired comparison matrix by its priority and sum the results:

$$0.31 \begin{bmatrix} 1 \\ 3 \\ 1/7 \end{bmatrix} + 0.62 \begin{bmatrix} 1/3 \\ 1 \\ 1/6 \end{bmatrix} + 0.06 \begin{bmatrix} 6 \\ 7 \\ 1 \end{bmatrix} = \begin{bmatrix} 0.94 \\ 1.97 \\ 0.20 \end{bmatrix}$$

4. Divide the results by the original priority vector to determine the relative importance or weight of each VEC:
   VEC A = 0.**94**/0.31 = **3.03**
   VEC B = 1.97/0.62 = 3.18
   VEC C = 0.20/0.06 = 3.33

5. The relative importances are used to interpret the relative significance of project impacts on each of the affected VECs.

SCOPE OF SIGNIFICANT IMPACTS

Strictly speaking, under the Canadian Environmental Assessment Act the determination of significance involves three general steps:

    i)   deciding whether project environmental effects are adverse;
    ii) deciding whether the adverse effects are significant;
    iii) deciding whether the significant adverse effects are likely.

However, only environmental effects as defined under section 2(1) of the Canadian Environmental Assessment Act can be considered in the determination of impact significance under federal EIA. Thus, according to the Canadian Environmental Assessment Agency's reference guide for impact significance, the determination of significance with regard to the impacts of a project can consider only:

- direct, adverse effects on the environment that are caused by the project;
- the effects of these environmental changes on health and socio-economic conditions, physical and cultural heritage, current traditional Aboriginal use of lands and resources, and structures, sites, or phenomena that are of historical, archaeological, palentological, or architectural significance;
- changes to the project caused by the environment.

There are two important issues here concerning the scope of significant impacts. First, under current EIA practice the majority of attention with regard to determining impact significance is focused on 'adverse' project effects. As a result, positive impacts are rarely treated with the same level of attention as negative impacts. This is not to say that positive impacts are necessarily more important than negatives ones, but if the objective of EIA is to make a positive contribution to sustainability, then increasing attention on creating and enhancing significant positive project impacts is required. Second, the socio-economic effects of a project may or may not be factors in determining the significance of project effects. For example:

If a socio-economic effect (such as job loss) is caused by a change in the environment (such as loss of fish habitat), which is in turn caused by the project, then the socio-economic effect *is* an environmental effect within the meaning of the Act and must be considered when determining significance and the related matters. If the socio-economic effect is not caused by a change in the environment, however, but by something else related to the project (for example, reallocation of funding as a result of the project), then the socio-economic effect is not an environmental effect within the meaning of the Act and cannot be considered in the determination of significance and the related matters. (Canada, 1994)

This appears to be inconsistent with the view that EIA provides an effective means of integrating environmental factors into planning and decision-making processes 'in a manner that promotes sustainable development' (Canada, 2003: preamble), if one accepts that sustainable development is based on the notion that human and ecological well-being are effectively interdependent.

KEY TERMS

acid mine drainage	rating
assimilative capacity	relative significance
fixed point scoring	residual impacts
impact significance	statistical significance
paired comparisons	

STUDY QUESTIONS AND EXERCISES

1. Determining the significance of environmental impacts is often a subjective process, particularly when issues of social concern are involved. Suppose, for example, that a proposed development is likely to result in the closure of a local outdoor public recreational area. How might you assess the impact of such an impact?

2. Obtain a completed project EIS from your local library or government registry, or access one on-line. Identify the types of criteria used to determine impact significance. Compare your findings to those of others. Are there noticeable similarities of significance criteria across impact statements? Is there evidence of 'sustainability criteria'?

3. Impact significance in EIA is often classified as 'major', 'moderate', 'minor', or 'negligible' without any qualification as to what these terms mean. Following the example illustrated in Figure 8.2, construct a simple impact significance matrix for the construction and operation of a waste incineration project. Identify project actions across the top and VECs down the side. Develop a legend for impact significance similar to that in Figure 8.2 and provide operational definitions for each level of significance based on the significance criteria discussed in this chapter.

4. Discuss the advantages of weighted impact assessment matrices over unweighted impact assessment matrices.

5. Assume that the following VECs have been identified for a large-scale energy project to be developed in your region:

   - air quality
   - employment
   - forests and vegetation
   - water quality
   - human health

   a) Use fixed-point scoring to assign weights to the affected VECs so that the total of all weights equals 1.

   b) Use a numerical rating scale to assign weights to the affected VECs so that 1 = not important, 3 = moderately important, 5 = important, 7 = very important, and 9 = extremely important.

   c) Construct a paired comparison matrix and, using the scale depicted in Box 8.1, calculate weights for each of the VECs.

   d) For each of the above approaches, standardize each of the weights for each VEC by using the following scaling parameter: $(i - i_{min}) / (i_{max} - i_{min})$, where i is the respective weight and i_{min} and i_{max} represent the minimum and maximum values of all weights respectively for the set of VECs. This will generate a standardized scale where the least important VEC = 0 and the most important VEC = 1.

e) Compare your results across weighting techniques? Are the weights different? Why?

f) Discuss the advantage of using paired comparisons over the other two weighting techniques.

6. Who should be responsible for determining impact significance? Are there certain criteria that would apply to practically all proposed developments in your area? What is the role of the publics in determining impact significance? Think back to the principles of public involvement discussed in Chapter 4.

REFERENCES

Broken Hill Proprieties Billiton (BHPB). 1998. *NWT Diamonds Project Environmental Impact Statement*. Vancouver: BHP Diamonds.

Canada. 1994. *Reference Guide: Determining Whether a Project Is Likely to Cause Significant Adverse Environmental Effects*. Ottawa: Canadian Environmental Assessment Agency.

———. 2003. Canadian Environmental Assessment Act. Assented to 23 June 1992. Ottawa: Queen's Printer.

Canter, L. 1996. *Environmental Impact Assessment*, 2nd edn. New York: McGraw-Hill.

Federal Environmental Assessment Review Office (FEARO). 1994. 'Reference Guide: Determining Whether a Project Is Likely to Cause Significant Adverse Environmental Effects', in Canadian Environmental Assessment Agency, ed., *The Canadian Environmental Assessment Act Responsible Authority's Guide*. Ottawa: Ministry of Supply and Services Canada.

Gibson, R.B. 2001. *Specification of Sustainability-based Environmental Assessment Decision Criteria and Implications for Determining 'Significance' in Environmental Assessment*. A report prepared under a contribution agreement with the Canadian Environmental Assessment Agency Research and Development Program. Hull, Que.: CEAA.

Hajkowicz, S.A., G.T. McDonald, and P.N. Smith. 'An Evaluation of Multiple Objective Decision Support Weighting Techniques in Natural Resource Management', *Journal of Environmental Planning and Management* 43, 4 (2000): 505–18.

Lawrence, D.P. 2004. *Significance in Environmental Assessment*. Research supported by the Canadian Environmental Assessment Agency's Research and Development Program for the Research and Development Monograph Series, 2000. Hull, Que.: CEAA.

Miller, G.A. 1956. 'The Magical Number Seven Plus or Minus Two: Some Limits on Our Capacity for Processing Information', *Psychological Review* 63: 81–97.

Saaty, T.L. 1977. 'A Scaling Method for Priorities in Hierarchical Structures', *Journal of Mathematical Psychology* 15: 243–81.

Sadler, B. 1996. *Environmental Assessment in a Changing World: Evaluating Practice to Improve Performance*. Final report of the International Study of the Effectiveness of Environmental Assessment. Fargo, ND: IAIA.

Shell Canada. 2002. *Environmental Impact Statement for the Jack Pine Mine, Phase I*. Calgary: Shell Canada.

Voisey's Bay Nickel Company. 1997. *Voisey's Bay Mine/Mill Project Environmental Impact Statement*. St John's: Voisey's Bay Nickel Company.

Wood, C. 1995. *Environmental Impact Assessment: A Comparative Review*. London: Longman Scientific and Technical.

Managing Project Impacts

IMPACT MANAGEMENT

Once potential impacts have been identified and their significance determined, the next phase of the EIA process is to design management strategies to address those impacts. The objective of any impact management system is to optimize the positive outcomes of a project and minimize the negative outcomes. The determination of potential impacts identified through the environmental assessment process is often imprecise because of uncertainties surrounding such factors as the project's design and timetable or because of exogenous factors. Thus, a management system designed to address central issues should be flexible to respond to unanticipated impacts as well as differences between the actual and predicted nature, level, or significance of the impacts.

Impact management involves those plans or strategies designed to avoid or alleviate anticipated impacts generally perceived to be undesirable and to generate or enhance effects seen as beneficial (Box 9.1). Canter et al. (1991), Sadar (1996), and Glasson et al. (1999) identify several types of strategies to address potentially negative impacts, including:

- avoidance;
- mitigation;
- rectification;
- compensation.

With respect to positive outcomes, impact management involves actions that will create new benefits and enhance existing benefits.

MANAGING NEGATIVE IMPACTS

Avoidance
Avoiding negative impacts at the outset is clearly the most desirable approach—if an impact can be avoided then the time and financial resources of impact mitigation or compensation are also avoided. Methods to avoid potentially adverse impacts can include such measures as setting regulatory standards concerning the use of toxic substances, scheduling project construction activities so they do not conflict with daily patterns of local socio-economic activity, land-use planning and zone designa-

---

Box 9.1   Potential Impacts and Management Approaches for a Road Construction Project

Component/Action	Impact	Management
dust	residential air quality worker health	water gravel surfaces require safety masks
habitat	vegetation destruction risk to wetlands	restore or create elsewhere avoid by rerouting
wildlife	habitat fragmentation vehicle collisions	create wildlife overpasses road signage
equipment hours	noise disruption emissions worker safety	schedule construction vehicle standards inspections and training
gravel infilling	stream sedimentation	sediment screens
economics	employment	hire locally

---

tion, and the construction of self-contained work camps to avoid potentially negative socio-economic effects that might be caused by site worker-community interaction. **Impact avoidance** does not necessarily enter the equation only after potential impacts are predicted and their significances determined, as much avoidance can occur early in the scoping process through identifying alternative locations, project designs, or implementation strategies. For example:

> *Activity:* road construction
> *Impact:* traffic congestion
> *Management:* schedule activity to avoid traffic rush hours

Mitigation

Not all potentially adverse impacts are recognized in advance. While mitigation is often used in a generic way to refer to impact management in general, strictly speaking **impact mitigation** is defined as the application of project design (Box 9.2), construction, operation, scheduling, and management principles and practices to *minimize* potentially adverse environmental impacts. For example, forest harvesting operations can lead to soil erosion and excessive runoff, which, in turn, may affect the quality of aquatic environments. The maintenance of **buffer zones**, or areas of undisturbed vegetation, is thus a desired impact mitigation practice in forestry. While buffer zones do not *avoid* soil erosion or surface runoff, they do reduce the impacts of erosion and runoff on aquatic environments. For example:

> *Activity:* forest clearing
> *Impact:* increased surface water turbidity due to erosion and runoff
> *Management:* establish stream buffer zones

Box 9.2 Mitigating Visual Impacts through Project Design

Declared a World Heritage Site by UNESCO in 1981, the Head-Smashed-In Buffalo Jump Interpretive Centre at Waterton Park, Alberta, is one of the largest and best preserved buffalo jumps in North America. Construction of an interpretative centre commenced on the site in 1986 and concluded near the end of that same year. A major concern for building architects and landscape planners was minimizing the visual impact of the interpretation centre on the bare, rolling plains of the prairie landscape. The building was subsequently designed to represent a harmonious relationship with the surrounding landscape. The building itself is totally submerged into the site, blending into the landscape both in design and colour. Visitors approach the main entrance to the building at the bottom of a small cliff, and proceed through an entrance flanked on both sides by retaining walls up to 10 metres in height, simulating bedrock scars depicting an archaeological dig site.

Rectification

Not all negative environmental impacts can be mitigated. In certain cases it is inevitable that environmental components will be temporarily damaged. **Rectifying impacts** refers to restoring environmental quality, rehabilitating certain environmental features, or restoring environmental components to varying degrees. For example, in cases where the construction phase of project development requires clearing of forested landscape, impact management efforts can focus on restoring the landscape post-construction to resemble the predisturbed state. In the case of a **sanitary landfill** site, for example, the site is typically returned to a vegetative state once the landfill is full and covered. While the rectified site condition may not exactly resemble the predisturbed condition, the objective is to return it to a more desirable condition than what was created by the project actions. For example:

> *Activity:* open pit mine development
> *Impact:* habitat loss due to site development
> *Management:* site reclamation with native vegetation following mine decommissioning

Compensation

Compensation for impacts is typically associated with residual impacts that remain after other impact mitigation or management options have been exhausted, or for which no management alternative exists. **Compensation** sometimes involves monetary or other benefit payments to those affected by the damage caused by the project, while in other cases it involves measures to recreate environmental habitats at an alternative site. In Canada, for example, a federal policy on aquatic-based habitat declares that there should be 'no net loss'. This is not to say that projects posing a threat to the destruction of aquatic habitat will not be approved; rather, any habitat that is lost must be compensated for. For example:

Activity: a mining operation requires drainage of a nearby lake for tailings disposal
Impact: loss of fish habitat and aquatic life
Management: proponent investment in a river or lake habitat restoration program
in the project region

Managing Positive Impacts

Creating Benefits

Impact management provides the necessary means not only to reduce or avoid potentially negative impacts, but also to create positive benefits from project development. The most desirable approach to managing positive impacts is to ensure that the project makes a positive contribution to the environment and society through the creation of new benefits, such as community economic growth or environmental improvement. For example:

Activity: hydroelectric dam construction
Impact: employment and economic expenditure
Management: creation of small business development support centre

Enhancing Benefits

Development projects often create as many positive impacts, particularly economic ones, as they do negative impacts. Indeed, if this were not case, democratic societies and their governments would be hard-pressed to justify the continued approval of such projects. Thus, an important management strategy is to enhance the benefits of potentially positive impacts and to maximize the duration of those impacts. This might involve, for example, ensuring the greatest possible distribution of financial benefits among affected communities over the greatest period of time. This was the case for the Voisey's Bay project, discussed in Chapter 1, which, for the first time, adopted an explicit sustainability mandate requiring the project proponent to enhance potentially positive impacts by committing to community infrastructure investment and worker training programs beyond the life of the mine project. Other common approaches include adopting 'buy local' and 'hire local' strategies. This was an important part of enhancing the benefits of the Hibernia offshore oil platform construction project at Bull Arm, Newfoundland, where Mobil Oil committed to ensuring that a minimum number of construction contracts would be let to local businesses and that a certain percentage of workers employed at the site would be Newfoundlanders. For the Ekati diamond mine in the Northwest Territories, similar impact enhancement strategies were adopted to maximize the benefits of project employment:

Activity: mine site construction and operation
Impact: local employment and income
Management: hire local; provision of financial counselling to employees

MEASURES THAT SUPPORT IMPACT MANAGEMENT

In addition to project-based impact management strategies, a number of broader measures facilitate management of the environmental impacts of industry operations. Perhaps the most internationally recognized measure is an environmental management system.

Environmental Management Systems

The International Organization for Standardization, commonly referred to as ISO, is a global federation of over 100 countries and is headquartered in Geneva, Switzerland. The organization was formed in 1947 to promote the development of international standards, primarily in product manufacturing. The first system for management standardization (ISO 9000 series) was introduced in the early 1990s. By the mid-1990s the ISO 14000 series was introduced to promote international industry standards in environmental management, and in 1996 the **ISO 14001** standardization for **environmental management system** (EMS) certification was approved. As of December 2003, 61,295 ISO 14001 certificates had been issued in 128 countries, with 1,242 certificates issued in Canada. EMSs are currently among the most widely recognized tools for managing the environmental affairs of industry.

An EMS is a voluntary industry-based management system by which an organization controls the activities, products, and processes that cause, or could cause, environmental impacts. By putting into effect an EMS a company seeks to minimize the environmental impacts of its operations (Box 9.3). EMSs emerged in response to industry's realization of the need for an integrated and proactive approach to managing industry environmental issues to assist in ensuring that it is complying with environmental regulations, adopting and following environmental objectives, and receiving economic benefits from improved environmental performance (Strachan et al., 2003). A well-designed EMS should help an organization to develop a proactive approach to environmental management, achieve a balanced view across all aspects of operations, enable effective and directed environmental goal-setting, and contribute to a more effective environmental auditing process. That being said, the link between meeting ISO standards and genuine improvement in environmental performance has not been clearly established. In fact, a study of toxic emissions from US automobile assembly facilities revealed that facilities with ISO 14001 EMS certification often fared worse than those without certification (Matthews et al., 2004).

Environmental Protection Plans

A second measure to ensure impact management effectiveness is an environmental protection plan (EPP). While an EMS is a voluntary industry initiative, an EPP is often a mandatory requirement in project-based EIA. As part of the EIA process, key impacts, issues, and management measures to address those impacts will have been identified. In Canada, if a project is approved then impact management measures may be further articulated in the form of an overall EPP designed to detail the specific impact management actions and the ways in which they are to be implemented.

Box 9.3   Environmental Management Systems

An EMS is ideally a cyclical process of continual improvement in which an organization is constantly reviewing and revising its management system. This consists of a 'plan-do-check-act' cycle, for which the first step is developing a management policy. Through policy development the organization's environmental goals and objectives are formulated, reflecting the important environmental aspects and targets as well as the legal requirements of the organization. This is followed by planning, implementation, monitoring, and program review. In this sense, not only can an EMS act as a regulatory system by which an organization seeks to meet industry standards, it can also be a potentially valuable tool for environmental management and improvement of industry opertions.

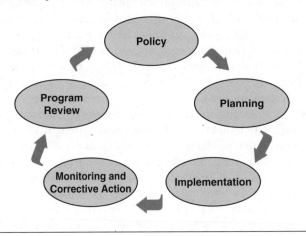

The particular components that make up an EPP are project-specific, and are custom-designed to fit the project context. EPPs may provide general information on management, construction, operation, and decommissioning of certain types of project, may be developed for different stages of the project, or can provide specific information concerning management techniques relevant to an individual project.

Impact Benefit Agreements

A third approach to facilitate impact management is an **impact benefit agreement** (IBA). IBAs are particularly useful for addressing the impacts generated by development projects on local communities, and are usually associated with compensation measures. IBAs are legally binding agreements between a proponent and a community that serve to ensure that communities have the capacity and resources required to maximize the potential positive benefits stemming from project development. This might include, for example, commitments to training programs, local hiring policies, or, in rare cases, regular cash payments. In Canada, IBAs are especially important in regard to industry gaining the co-operation of Aboriginal communities and for these communities to achieve certain guaranteed benefits (Box 9.4).

Box 9.4 Falconbridge Raglan Agreement

The Raglan property, located in the Nunavik territory of northern Quebec, stretches approximately 55 kilometres from east to west and contains many high-grade ore deposits, including nickel and copper. The reserve consists of a proven 8.3 million tonnes, grading 2.86 per cent nickel and 0.77 per cent copper. Falconbridge, an international producer of nickel, copper, cobalt, and platinum group metals, commenced production at Raglan in 1998. The project consists of open pits, an underground mine, a concentrator, a power plant, and housing and administrative buildings, and employs approximately 470 people. The mine site is linked by all-weather roads to storage and shipping facilities at Deception Bay. The James Bay and Northern Quebec Agreement does not require negotiation of IBAS and the project is not located on Inuit-owned lands. However, Falconbridge needed the shoreline for shipping and was concerned that an Inuit offshore claim recognized by the federal government could affect the project. Thus, in 1995 Falconbridge signed an agreement with the Makivik Corporation that includes profit-sharing and guaranteed contributions to the Inuit people of Nunavik and formed a basis for the IBA. As part of the IBA, the Inuit would receive $14 million plus 4.5 per cent of mine profits, estimated at $60 million over 15 years, have a representative appointed to Falconbridge mine's board of directors, and form a joint committee to oversee training programs. In addition, Inuit companies and communities would be given preference for contracts with the mine.

Source: O'Reilly and Eacott (1999). See also <www.falconbridge.com>.

EIA AND IMPACT MANAGEMENT IN CANADA

EIA is more than just a tool to identify and predict potential environmental impacts; it should also play an important environmental management role. Impact management must be planned in an integrated and coherent fashion to ensure that such measures are effective and non-contradictory and that they do not shift the problem from one environmental component or sector of society to the next (Glasson et al., 1999). Thus, impact management is not limited to any one point in the EIA process. From the outset of project design and scoping, measures should be considered to avoid potentially adverse impacts and to create or enhance positive ones.

However, under the Canadian Environmental Assessment Act impact management (referred to only as mitigation under the Act) intends to reduce, eliminate, or to make less severe the adverse environmental effects of a project. Storey and Noble (2004) note that while most biophysical effects arising from human activities are adverse, many social, economic, and other human effects are not. Increased employment, training, and business development, for example, are usually regarded as positive outcomes. Mitigation, strictly defined, means to 'make less severe'. While important, it may be only one of several types of action that those responsible for managing project effects may need or wish to take. There is, however, a tendency to use mitigation as a generic term to cover all types of management. This, together with the emphasis of the Act on adverse biophysical effects, reinforces a narrowness of thinking about the consequences of project developments. The emphasis on simply reducing the severity of effects reflects a 'damage control' perspective rather than a proactive one in which actions taken might conceivably have a positive effect. If sustainable development is truly a goal of Canadian EIA, then there is a need to adopt a broader perspective on management strategies for all types of potential effects.

Rather than simply aiming to mitigate adverse effects, managers should consider a full range of management options, and the language of Canadian EIA should reflect this. As a draft guide to good practice, Mitchell (1997) suggests a hierarchy of impact management practices, notably:

- avoiding impacts at the source;
- reducing impacts at the source;
- mitigating impacts on site;
- mitigating impacts on the environmental receptor;
- repairing impacts;
- compensating for impacts;
- enhancing positive impacts.

From a broader sustainability perspective, it might be argued that at the top of this hierarchy should be the creation and enhancement of overall, long-term positive benefits to environment and society and recognition of the trade-offs involved in doing so. That being said, mitigation measures are of little value unless they are actually implemented, monitored for effectiveness, and adjusted accordingly.

KEY TERMS

buffer zone
compensation
environmental management system
environmental protection plans
impact avoidance

impact benefit agreement
impact mitigation
ISO 14001
rectifying impacts
sanitary landfill

STUDY QUESTIONS AND EXERCISES

1. Suppose you are responsible for negotiating an IBA for a small community (population less than 5,000) about to be the recipient of a large mining operation. What items might you want to negotiate with the proponent and the government for inclusion in the IBA?
2. Assume a simple highway construction project through a small community. Develop a list of project activities and potential impacts, and propose as many different types of impact management measures as possible, from avoidance to enhancement.
3. Obtain a completed project EIS from your local library or government registry, or access one on-line. Document the nature of impact management measures. For example, are most impact management measures based on avoidance, mitigation, rectification, or compensation? Are there any impact management measures that emphasize creating or enhancing positive project impacts? Generate a list and compare your results with those of others.

4. Visit the International Organization for Standardization Web site at <www.iso.org>.
   a) How many countries were ISO members this past year?
   b) Is your country an ISO member?
   c) What does ISO identify as the principal benefits to businesses of ISO 14001 certification?

REFERENCES

Canter, L.W., et al. 1991. 'Identification and Evaluation of Biological Impact Mitigation Measures', *Journal of Environmental Management* 33: 33–5.

Glasson, J., R. Therivel, and A. Chadwick. 1999. *Introduction to Environmental Impact Assessment: Principles and Procedures, Process, Practice and Prospects*, 2nd edn. London: University College London Press.

Matthews, D.H., G.C. Christini, and C.T. Hendrickson. 2004. 'Five Elements for Organizational Decision-making with an Environmental Management System', *Environmental Science and Technology* 38, 7: 1927–32.

Mitchell, B. 1997. *Resource and Environmental Management*. Harlow, England: Addison Wesley Longman.

O'Reilly, K., and E. Eacott. 1999. 'Aboriginal Peoples and Impact and Benefit Agreement: Summary of the Report of a National Workshop', *Northern Perspectives* 24, 4. Available at: <www.carc.org/pubs/v25no4/index.html>.

Sadar, H. 1996. *Environmental Impact Assessment*, 2nd edn. Ottawa: Carleton University Press.

Storey, K., and B. Noble. 2004. *Toward Increasing the Utility of Follow-up in Canadian EA: A Review of Concepts, Requirements and Experience*. Report prepared for the Canadian Environmental Assessment Agency. Hull, Que.: CEAA.

Strachan, P.A., I. McKay, and D. Lal. 2003. 'Managing ISO 14001 Implementation in the United Kingdom Continental Shelf (UKCS)', *Corporate Social Responsibility and Environmental Management* 10: 50–63.

Post-Decision Monitoring

FOLLOW-UP

Much of what has been covered in this manual has focused on what happens before a decision is made to approve or not to approve a project. This has typically been the focus of EIA—a 'build it and forget about it' syndrome. In this way, EIA is no more than a linear, predictive process with no mechanism to ensure quality assurance of the process itself, verify impact predictions, and evaluate the effectiveness of measures proposed to manage actual project impacts. Only in recent years has attention been given to the follow-up stage of EIA—that which happens after a decision is made—but still only limited requirements exist in regard to undertaking post-decision analysis (Box 10.1).

In essence, follow-up is the element that can transform EIA from a static to a dynamic process and the missing link between EIA and effective **life-cycle assessment** (Arts et al., 2001). The rationale for follow-up is similar to that of EIA itself—trying to come to grips with the uncertainties intrinsic to a particular activity (ibid.). Some EIA practitioners interpret follow-up to mean strictly ensuring that mitigation measures identified in the assessment were implemented, while others view follow-up as an umbrella that encompasses the activities, such as routine monitoring or auditing, undertaken during the post-decision stages of the environmental assessment process.

Box 10.1 Requirements for EIA Follow-up

Australia: Discretionary provisions for follow-up, not common practice.

Canada: Provisions under Canadian Environmental Assessment Act for verifying impact predictions and mitigation effectiveness, but no mechanisms to ensure compliance or to outline responsibilities.

Netherlands: Specific requirements for comparison of monitoring data to the EIS, but rarely carried out in practice.

United Kingdom: No legislative EIA requirement for monitoring project effects; monitoring takes place under various legislation or planning regulations, which are not directly related to the project EIA itself.

United States: Monitoring is discretionary, except where specific requirements in the EIA decision state that mitigation outcomes must be verified and reported.

In Canada, under the Canadian Environmental Assessment Act, a follow-up program means a program for (1) verifying the accuracy of the environmental assessment of a project, and (2) determining the effectiveness of any measures taken to mitigate the adverse environmental effects of the project. One question that emerges from this is the value of verifying the accuracy of impact predictions when project environmental conditions are constantly changing. In this sense, it is perhaps the second requirement of follow-up, that of a management function, that is of greater value. This latter perspective is adopted in this chapter.

Follow-up Components

Environmental impact assessment follow-up can be thought of as consisting of three interrelated components: monitoring, auditing, and ex-post evaluation. Generally speaking, **monitoring** is an activity designed to identify the nature and cause of change. More specifically, it is a data collection activity undertaken to provide specific information on the characteristics and functioning of environmental and social variables (Bisset and Tomlinson, 1988). Monitoring usually consists of a program of repetitive observation, measurement, and recording over a period of time for a defined purpose (Arts and Nooteboom, 1999), the objective of which is to detect if change in a particular variable has occurred and to estimate its magnitude. Determining the causes of change and whether such changes are the consequence of the project is an essential part of monitoring.

Originating in economics and accountancy, **auditing** refers to objective examination or a comparison of observations with predetermined criteria. There are a variety of types of audits (Box 10.2), but auditing generally is a periodic activity that involves comparing monitoring observations with a set of criteria, such as standards or expectations, and reporting the results. Whereas monitoring is often a frequent or continual process, auditing is usually a periodic or single event. There is little point

---

Box 10.2 Types of EIA Audits

Draft EIS audit: Review of the project EIS according to its terms of reference.

Project impact audit: Determination of whether the actual environmental impacts of the project were those that were predicted.

Decision-point audit: Examination of role and effectiveness of the EIS based on whether the project is allowed to proceed and under what conditions.

Implementation audit: Determination of whether the recommendations included in the EIS were actually implemented.

Performance audit: Examination of the proponent's environmental management performance and ability to respond to environmental incidents.

Predictive technique audit: Comparison between actual and predicted effects of the project.

Source: Tomlinson and Atkinson (1987).

in collecting monitoring data unless those data are subject to some form of comparative analysis or audit (Arts and Nooteboom, 1999).

A third and related concept, used principally by the Dutch, is that of **ex-post evaluation**, which refers to the collection, structuring, analysis, and appraisal of information concerning project impacts, and making decisions on remedial actions and communication of the results of this process (Arts, 1998: 75). The differences and relationships among these three components are illustrated in Figure 10.1.

RATIONALE FOR POST-DECISION MONITORING

There are several, often overlapping, reasons for undertaking monitoring activities. However, three broad reasons for post-decision-monitoring can be identified: monitoring for compliance, monitoring progress, and monitoring for understanding.

Compliance Monitoring

The primary purpose of **compliance monitoring** is to determine project compliance with regulations, agreements, or legislation. In this sense follow-up has a *control* function, to ensure that a project is operating within specified guidelines. There are several types of compliance monitoring, including inspection monitoring, regulatory permit monitoring, and agreement monitoring.

Inspection monitoring, the simplest form of monitoring, is site-specific and involves checking to ensure that operating procedures are being followed and that environmental degradation is not occurring. Inspection monitoring typically involves on-site visits and regular reporting of relevant activities and is mostly used for regular checking for compliance with agreed-upon procedures and operation within acceptable standards of safety.

Regulatory permit monitoring is also site-specific and involves the regular documentation of conditions required for maintenance or renewal of a permit, such as permits for the operation of waste disposal systems.

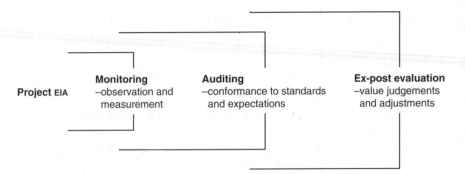

	Monitoring	Auditing	Ex-post evaluation
Project EIA	–observation and measurement	–conformance to standards and expectations	–value judgements and adjustments

Figure 10.1 Follow-up Components

Source: Based on Arts (1998).

The **monitoring of agreements** between project proponents and affected groups, such as impact benefit agreements, are becoming commonplace, particularly in Canada. These sometimes include a monitoring component to track changes in population, housing, and other infrastructure demands in order to assign costs associated with the project and to ensure compliance with stated commitments.

Progress Monitoring

The purpose of progress monitoring is to confirm anticipated outcomes and to alert managers to unanticipated outcomes. In this sense follow-up has a *watchdog* function and allows managers to measure project and environmental progress and to respond to adverse environmental change in a timely fashion when necessary. Common approaches to progress monitoring include ambient environmental quality monitoring, monitoring for management, cumulative effects monitoring, and project evaluation monitoring.

Ambient environmental quality monitoring is concerned with the effects of the project on its surrounding environment. Information collected prior to project approval and implementation at and near the site and at control sites can provide baseline data against which to compare data collected during the development, operation, and post-project phases. While most ambient environmental quality monitoring is associated with measures of the status of biophysical phenomena, such as air and water quality, socio-economic changes such as quality of life or health and well-being can also be the focus.

Monitoring for management may include tracking and evaluating changes in a range of environmental, economic, and social variables. This type of monitoring is usually associated with high-profile projects with uncertain outcomes, which have the potential to result in significant adverse outcomes unless prompt action is taken to address issues as they emerge. The size and scope of the monitoring requirements may be such that the co-ordination of the monitoring program is undertaken by a formally constituted and funded monitoring organization, such as the Independent Environmental Monitoring Agency (IEMA) established to oversee outcomes from the Ekati Diamond Project in northern Canada (www.monitoringagency.net) and to ensure that the project meets the requirements set forth in IEMA's environmental approval. Monitoring for management is the primary focus of its mandate, and finding solutions to environmental management issues arising from the project is a primary objective.

Cumulative effects monitoring is less site- and project-specific and attempts to monitor the accumulated effects of developments within a particular region. The broad range of interests involved and the need for co-ordination mean that cumulative effects monitoring is best achieved by an organization mandated with monitoring responsibilities. The nature of cumulative environmental effects is discussed in greater detail in the next chapter.

Also referred to as productivity measurement, **project evaluation monitoring** is concerned with measuring a project's performance in respect to established goals or objectives, such as overall efficiency. The focus is often social or economic in nature, and includes measures of performance of programs designed to provide job training to address social concerns.

Monitoring for Understanding

A third reason for monitoring is to better understand the complex relationships between human actions and environmental and social systems. In this sense, follow-up has a *learning* function by increasing knowledge and understanding that can be applied to the science of assessment of future projects or related policy decisions. Monitoring for understanding includes experimental monitoring and monitoring for knowledge.

The purpose of **experimental monitoring** is to generate information and knowledge about environmental systems and their impacts through research methodologies that test specific hypotheses. Whereas the above approaches, while designed to monitor changes, are not guided by any anticipated consequences, experimental monitoring is guided by questions to test specific hypotheses. In this sense, experimental monitoring is science-driven rather than motivated by impact management per se.

Monitoring for knowledge is a type of data collection and reporting that often takes place well after impacts occur. Rather than being used for impact management purposes, the data are used to provide insights for the management of future projects (Box 10.3).

Box 10.3   Monitoring for Knowledge: The Alma Smelter, Quebec

The Alma smelter began operation in 2001 in Alma, Quebec. It employs 865 people and has a capacity of 407,000 metric tons of aluminum ingots. Researchers at the University of Quebec at Chicoutimi undertook a five-year study between 1998 and 2003 to document the social impacts of the project on Alma, which has a population of 30,126. The study was carried out in three parts in 1998, 2000, and 2002, corresponding to the planning, construction, and operating phases of the project. The study examined quality of life of local residents using a series of subjective indicators to reflect perceptions and degree of satisfaction with living conditions.

The stated medium- and long-term research objectives of the research were to:

- develop an impact follow-up model of Alma's aluminum smelter before 2001;
- generate applied knowledge concerning:
  - economic characteristics of study zones;
  - citizen perceptions concerning quality of life and the environment;
  - expected social and economic impacts before, during, and after construction of the smelter;
  - assessment methods and impact follow-up models;
  - processes, modes, and analysis of public and community participation.
- set up a geo-referenced database for urban and environmental planning and management;

Continued

- propose, if needed, preventive and mitigation measures during the plant's set-up;
- generalize results in regard to the establishment of future electrolysis plants in Quebec;
- allow development of appropriate expertise in the field of environmental monitoring and regional development.

Quality of life, based on a related concept of health proposed by the World Health Organization, was defined as 'a state of complete physical, psychological, and social well-being'. The approach used was to measure a series of subjective indicators, reflecting residents' perceptions and degree of satisfaction with living conditions, through face-to-face interviews with members of the community. The same questionnaire was used in each of the three phases of the study to survey a stratified sample of the population by census enumeration area. Response rates ranged between 61 per cent and 73 per cent.

The study indicated that, as stakeholders, residents were generally satisfied, despite an increase in the impacts during the construction phase of the smelter. Responses gathered showed a broad consensus on a number of issues. An Alma resident is generally someone who:

- is satisfied with his/her quality of life with respect to the biophysical and community environment as well as the economy;
- is concerned about issues related to health and family;
- sees Alma as a dynamic community and does not plan to move;
- perceives few environmental risks, while remaining aware of the possible effects of industrial activity;
- is somewhat dissatisfied with the traffic situation and noise caused by blasting during the construction phase.

Follow-up for impact management purposes did not appear to be a primary focus of the research and there is no information to suggest that preventive or mitigative measures were identified, proposed, or acted on with respect to the project. Rather, the emphasis was on using follow-up as a database and learning tool, information and experience from which could be applied to other, similar projects.

Source: From Storey and Noble (2004). For additional details, see the research project Web site at <www.uquac.ca/~msiaa>.

REQUIREMENTS FOR EFFECTIVE MONITORING

In Canada, and elsewhere, post-decision monitoring has been described as less than satisfactory. While weak or non-existent legal requirements and institutional support mechanisms may, in part, be contributing to the current state of practice, a number of substantive and procedural elements are also required to facilitate effective post-decision EIA monitoring programs.

Objectives and Priorities Identified

Perhaps the most important requirement for effective monitoring is the clear articulation and identification of follow-up and monitoring program objectives and priorities at the outset of monitoring program design. If a clear set of objectives for monitoring does not exist, whether for monitoring for compliance, performance, or knowledge, then one cannot sufficiently determine if the monitoring program itself is effective. The specific objectives of monitoring programs vary from one project to the next (Box 10.4) and may include such objectives as verifying impact predictions or ensuring regulatory compliance for project licensing. Failure to state clear program objectives and priorities leads to dissatisfaction with and confusion regarding that which the monitoring program is actually trying to achieve. Consideration should be given to monitoring objectives and priorities from the outset of EIA, during the scoping process when key environmental issues are first identified, as this will allow pre-development baseline monitoring. Specific monitoring program design can be delayed until the post-decision stage when project design elements are finalized and priorities for management purposes can be better determined.

Targeted Approach to Data Collection

An effective monitoring program requires that specific variables and indicators are identified for which data will be collected. Such components must be comparable over space and time. In order to differentiate project-induced change from natural change, these variables and indicators and the data derived from them must be comparable to previous, current, and forecasted baseline conditions. One approach is to focus on early warning indicators.

Measurable parameters of either a non-biological or biological nature, **early warning indicators** serve to indicate stress on particular VECs before those VECs themselves are adversely affected. Examples of early warning indicators might include water quality, sediment quality, or biological indicators for assessing early stress on marine environments, or worker productivity and number of sick days

---

Box 10.4 Follow-up Program Objectives for Two Canadian Mining Projects

Voisey's Bay nickel mine-mill project

1. To continue to provide baseline data so that project activities can be scheduled or planned to avoid or reduce conflict with VECS.
2. To verify earlier predictions and evaluate the effectiveness of mitigation to lower uncertainty or risk.
3. To identify unforeseen environmental effects.
4. To provide an early warning of undesirable change in the environment.
5. To improve understanding of environmental cause-and-effect relationships.

Ekati diamond mine project

1. To ensure regulatory compliance.
2. To measure operational performance and the effectiveness of mitigation strategies.
3. To monitor natural environmental changes as well as those caused by the project (environmental effects monitoring).
4. To assess the validity of impact predictions.
5. To trigger response to and mitigation of unexpected adverse effects.

taken as early warning indicators of worker stress (Table 10.1). Early warning indicators must be:

- directly or indirectly related to the VEC;
- physically possible to monitor;
- amenable to quantitative analysis.

Hypothesis-based or Threshold-based Approaches

For each variable to be monitored, significance levels and probability levels must be specified as part of the monitoring program protocol. Successful impact monitoring requires that such information be formulated as testable hypotheses; analyses of significance will have meaning only when they are made against an a priori null hypothesis (Bisset and Tomlinson, 1988). For each indicator, appropriate thresholds should

Table 10.1 Selected Environmental Components, VECs, and Monitoring Indicators for the Cold Lake Oil Sands Project

Environmental component	Issues of concern	Valued components	Warning indicators
Air systems	Acidic deposition, odours, greenhouse gas emissions	Air quality	Emitted gases transported over long distances (NOx, SO_2)
Surface water	Lowering of lake water levels, contamination of water	Water quality and quantity	Combined water volume withdrawals, water quality constituents affecting drinking water standards
Groundwater	Depletion of aquifers	Potable well water	Combined water volume withdrawals
Aquatic Resources	Contamination of fish, increased harvest pressures	Sport fish species	Northern pike populations/health
Vegetation	Loss of vegetation through land clearing, effects of airborne deposition	Vegetation ecosites	Low bush cranberry, aspen, white spruce
Wildlife	Loss, sensory alienation and fragmentation of habitat, direct mortality due to increased traffic and hunting harvest	Hunted and trapped species	Moose, black bear, lynx, fisher populations/health

Source: Based on Hegmann et al. (1999).

thus be established as a measure of environmental effects (Table 10.2). Such thresholds may be established based on:

- consultation with regulatory agencies;
- levels of acceptable change;
- scientific research or recommendation;
- project goals or objectives.

This approach, which is common to many biophysical EEM programs, is often based on testing to determine if changes in specific indicators exceed stated threshold levels. Monitoring of these indicators helps to verify that the management measures implemented are effective, but since there is no attempt to determine the specific level of the effect, predictive accuracy is not the primary focus of the assessment (Storey and Noble, 2004). Such an approach was adopted for the Hibernia offshore oil platform construction project in Newfoundland (Box 10.5).

Control Sites
Where possible, **control sites** should be established for reference monitoring locations to compare with the treatment (project-affected) locations. This will assist in differentiating correctly between project impacts and natural change and will facilitate spatial and temporal consistency of the effects monitoring program with initial project impact predictions. In cases where a well-established control site is not readily available, it may be possible to create an artificial one using a **gradient-to-background approach**.

Table 10.2 Selected Discharge Thresholds and Monitoring Requirements for Kerr-McGee Offshore Oil and Gas Exploration and Development, Gulf of Mexico

Parameter	Threshold	Monitoring Frequency	Monitoring Approach
Oil and grease	42 mg/l daily maximum 29 mg/l monthly average	once per month record daily maximum	Grab sample at effluent source
Free oil	no free oil	estimate monthly average	Visual sheen
Toxicity	7-day minimum no observable effect concentration (NOEC) and monthly average minimum NOEC	record lowest NOEC	Grab sample at effluent source

Source: MMS (2001).

Box 10.5   Hibernia Offshore Platform Construction Project, Biophysical Environmental Effects Monitoring (BEEM) Program

The Hibernia offshore oil field was discovered on the Grand Banks of Newfoundland in 1979. Its proposed development was subject to a panel review under the Canadian Federal Environmental Assessment Review (FEARO) process. Approval for the project was granted in 1986 and development began in 1990. The Hibernia Management and Development Company (HMDC), in conjunction with relevant agencies forming part of the Hibernia Construction Sites Environmental Management Committee, established a multi-year (1991–6) program to monitor the effects of the Hibernia Gravity Base Structure (GBS) construction site on the marine environment of Bull Arm, Trinity Bay. Survey data on benthic fauna, marine fish and shellfish habitat, chemical and biological characteristics of the water column, sediments and resident benthic fauna, and fisheries utilization had been collected as part of the assessment process, which provided both a baseline for subsequent monitoring activity and information for determining various monitoring criteria.

The construction site at Great Mosquito Cove, Newfoundland, and the adjacent deepwater site were considered the major potential point sources of effluent and/or accidental discharge. The area is rich in marine invertebrates, fish, and marine mammals, and Bull Arm had traditionally supported a diverse inshore commercial fishery, which formed the economic base of many of the communities in the area. The general objectives of the biophysical environmental effects monitoring (BEEM) program included an assessment of the effectiveness of environmental protection and mitigation measures, but also included providing early warning of undesirable change and assurance that impacts predicted to be insignificant in fact were insignificant ('comfort monitoring').

Selection of monitoring variables to reflect significance to the environment and management needs was based on the following criteria (Buchanan et al., 1990): (1) social, economic, or biological importance; (2) potential that the variable would be affected by the project; (3) species are common/accessible in the area; (4) species are amenable to scientific sampling and monitoring; (5) baseline data are either not required or easily collected; (6) variables can be monitored over time for relatively low cost.

Variables chosen to reflect the specific objectives of the program included: (1) mixed function oxygenase (MFO) activity and bile metabolite level in winter flounder; (2) gill deformity in winter flounder; (3) trace metal and hydrocarbon levels in blue mussel tissue; (4) trace metal and hydrocarbon levels in shallow and deep water sediments; and (5) near-shore sedimentation rates (LGL, 1993).

For each measured variable an impact (null) hypothesis was developed, stating that: 'activities associated with site development and construction of the GBS and Topsides in Bull Arm will not elevate the concentration or degree of the variable to a level which exceeds the maximum allowable effects level (MAEL) for that variable.' The MAELs for the variables were based on values from different sources. For example, those applying to mussels and sediments included the Canadian Fish Health Inspection/Food and Drug Act Regulations, the Food and Agriculture Organization of the United Nations, CEPA Ocean Dumping Regulations, and Canadian Marine Environmental Quality Guidelines. For variables without values from these sources, other comparable local data and professional opinion were used to define MAELs (ibid.).

Continued

> Field samples of mussels, flounder, and sediments were collected and prepared for analysis by consultants to HMDC. Established commercial and university laboratories undertook analysis of the samples and the results summarized and reported annually (ibid., 1993–7). The findings from the analyses were that all of the null hypotheses developed for the BEEM program could not be rejected. In other words, the construction project did not have impacts on the marine environment beyond acceptable levels. A similar approach was adopted for the offshore environmental effects monitoring program for the offshore development phase of the Hibernia project and subsequent offshore developments in the area—Terra Nova (2002) and White Rose (2005)—have adopted similar monitoring programs.
>
> Source: Storey and Noble (2004).

This approach assumes that there is a well-defined, localized source of impact, such as pollution, and that effects can be monitored at increasing distances from the point source. With increasing distance from the point source, effects should decrease and eventually reach background or ambient conditions. The point on this gradient at which background levels are attained is considered the control site (Figure 10.2).

Continuity

There should be continuity in data collection and handling procedures to ensure that monitoring data are transferable and comparable. Failure to achieve continuity may lead to problems in comparing monitoring results and the significance of those results from one monitoring period to the next. Such quality control problems in the continuity of data collection were evident in the Rabbit Lake uranium mine project in Saskatchewan. After decades of biophysical monitoring and data collection, the proponent was unable to make a direct connection between project actions and the impacts of project-induced environmental change (Box 10.6).

Adaptability, Flexibility, and Timeliness

Care should be taken to ensure that as project or environmental conditions change, data quality are consistent and comparable over time. While objectives and indicators need to be established at the outset, project changes, unanticipated effects, emerging project concerns, and indicator choices may warrant changes in elements of the monitoring as it proceeds (Davies and Sadler, 1990). This requires that monitoring and

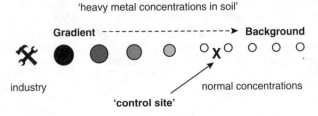

Figure 10.2 Gradient-to-Background Approach

Box 10.6 Continuity in Monitoring Data: The Rabbit Lake Uranium Mining Project

Cameco's Rabbit Lake mining project located in northern Saskatchewan is the province's oldest operating uranium mining and milling facility. Since its discovery 1968, the Athabasca Basin of northern Saskatchewan continues to be the world's premier exploration region for high-grade uranium deposits. In 1987 subsequent exploration activity of the Rabbit Lake mine site identified several additional radioactive occurrences in the area, and the proponent, Cameco Corporation, submitted an EIS to federal and provincial regulatory agencies for approval to mine three new ore bodies. The project was approved and a licence issued for development.

Four years later, a joint federal-provincial EIA panel was appointed to examine the environmental, health, and socio-economic effects of uranium mining activities in northern Saskatchewan, including the Rabbit Lake mine and extension. The panel's report, released in 1993, identified contamination of the biophysical environment and subsequent exposure pathways to radionuclides and heavy metals of primary concern. In its presentation to the panel, Cameco noted that it had collected baseline data and monitored the biophysical environment near the mine site since the late 1970s. However, the panel noted that while monitoring requirements were in place that met regulatory requirements and data collection was ongoing, there was some concern over the quality of the monitoring data, consistency of methods used to test for radionuclides and trace elements, and effectiveness of the monitoring program in determining the impacts of mining activities.

Monitoring and testing procedures had changed several times throughout the 1980s, and data collected during 1989 and 1990 were discarded due to quality problems. After more than a decade of biophysical monitoring and data collection, there were few comparable data concerning the effects of radionuclides from mining operations on fish— a valuable resource base to northern residents.

---

reporting must be completed in a timely fashion to allow those using the results to make prompt responses. While data collected late in the process are still valuable for reflection and learning, such monitoring processes are often a waste of time from the perspective of impact management.

Inclusiveness

Follow-up programs should address not only traditional biophysical impacts, but also socio-economic and other human and cultural impacts. In practice, however, social issues have received less than sufficient attention in post-decision monitoring. The result is that while the assessment of socio-economic and other effects on the human environment are often given as much attention as biophysical effects in impact statements and panel reports, this attention is less likely to carry over into follow-up—particularly for the 'social' components of socio-economic impacts (Box 10.7). When social and other human issues do carry over to the follow-up stage, they are treated with considerably less rigour than biophysical components (Box 10.8). Proponents often wish to distance themselves from socio-economic follow-up, due in part to the complexity of the pathways that link project actions to socio-economic impacts. In the case of the Cluff Lake uranium mine in northern Saskatchewan, the

Box 10.7   Canadian Experiences with Socio-economic Monitoring: Confederation Bridge Project

In May 1989 Public Works Canada submitted a Bridge Concept Assessment document for review. The proposed project was the construction of a 13-km bridge over the Northumberland Strait, extending from Cape Tormentine, New Brunswick, to Borden, Prince Edward Island. The mandate of the EA panel reviewing the proposal was to examine the environmental and socio-economic effects, both beneficial and harmful, of the proposed bridge. There was consensus on the need for improved transport service between Prince Edward Island and New Brunswick, but the panel's recommendation was that the risk of harmful effects, particularly concerning delays in spring ice retreat, risk of damage to the ecosystem, and the displacement of more than 600 Marine Atlantic ferry workers, was such that the project should not proceed.

The panel's recommendation was not accepted by the federal government, and following a 1990 study by an independent Ice Committee the Minister of Environment approved the project based on the committee's conclusion that the designs for the bridge under consideration would meet the panel's ice-out criteria. Construction begun in the fall of 1993 and the Confederation Bridge was completed in the spring of 1997. An Environmental Management Plan (1993) was prepared, included in which was a long-term Environmental Effects Monitoring (EEM) Program. In line with the CEAA definition of follow-up and the emphasis on the physical environment, the goals of the EEM were to evaluate the effectiveness of environmental protection procedures and to verify predictions regarding the potential biophysical effects of the project. Ice characteristics were initially documented in 1993, before the bridge was constructed, studied throughout project construction, and monitored after the bridge was complete. Notwithstanding the concern expressed in the initial panel report over the issue of displacement of ferry workers, there was no formal attempt to follow up on this or any other of the potential social or economic effects raised in the assessment document, at the public hearings, and by the panel.

Source: Storey and Noble (2004).

Box 10.8   Canadian Experiences with Socio-economic Monitoring: Ekati Diamond Mine

In 1994 the Canadian Department of Indian and Northern Affairs and Northern Development initiated an environmental review of Canada's first diamond mine, 300 km northeast of Yellowknife in the Northwest Territories. The proposal involved the development of a diamond mine in an area of unsettled and overlapping Aboriginal land claims where there had been little previous industrial development. The proponent, now BHP Billiton (BHPB), submitted its assessment documents in 1994 and a full panel review followed.

The importance of follow-up was recognized and a variety of suggestions were made to ensure that this was undertaken. The panel, based on information from the government of the NWT (GNWT) that it had already identified and collected data on a number of health and wellness indicators, recommended a partnership approach, primarily between the proponent and the GNWT.

Continued

In 1996 a Socio-economic Agreement was signed between the GNWT and BHPB for the Ekati project. The Agreement is intended to promote the development and well-being of the people of the NWT, particularly people in communities close to the mine. The Agreement focuses on monitoring and promoting social, cultural, and economic well-being. The GNWT is responsible for the establishment and maintenance of the monitoring program. The objective is to look for ways to strengthen opportunities and to mitigate negative effects associated with the project. Public statistics and surveys of mine employees are the primary data sources used to track a series of indicators chosen to reasonably match the possible effects identified during the assessment phase for the project. In addition to data on social stability and community wellness, BHPB issues its own reports describing realization of business and employment opportunities and also reports directly to the Aboriginal groups with whom it has signed impact benefits agreements.

Public statistics for 14 indicators are currently used to monitor and assess the effects of the Ekati project. Data are monitored and reported for the small NWT communities in the West Kitikmeot Slave area, Łutselk'e, Rae-Edzo, Rae Lakes, Wha Ti, Wekweti, Detah, and Ndilo, as well as for Yellowknife. These data are then compared with information for the rest of the NWT. The results are presented in an annual report. While these data provide indications of overall changes within the region for the indicators tracked, most of the results are at best inconclusive in linking social and economic changes in the communities to the diamond project activities. Cause and effect are always difficult to determine for social variables, but the 'coarseness' of the indicators used to measure change, the small size of some communities, and the quality of some of the available data make it particularly difficult to do so in this case.

Key socio-economic monitoring indicators for the BHPB Ekati project include:

Social stability and community wellness
– number of injuries
– number of potential years of life lost
– number of suicides
– number of teen births
– number of children in care
– number of complaints of family violence
– number of alcohol/drug-related crimes
– number of property crimes
– number of communicable diseases
– housing indicators

Non-traditional Economy
– average income of residents
– employment levels and participation
– number of income assistance cases
– high school completion
– cultural well-being

Sources: GNWT (2001); Storey and Noble (2004).

final panel report of the Cluff Lake Board of Inquiry (1978: 174) explicitly recognized the difficulty of assessing the social and other human impacts associated with uranium mining activities, stating:

there now exists in the north (and it has nothing to do with uranium mining) a social disorder. . . . To superimpose upon that kind of society a project such as a uranium mine and mill which has the potential of exacting additional social costs and then try and measure those additional costs presents a near impossible task.

In principle, according to discussions with an industry representative, there are often too many confounding factors in the communities to ever be able to tell whether or not there was an effect, because in many cases communities are too far away for monitoring direct effects based on scientific risk and pathways monitoring.

If sustainable development is the objective of EIA, the biophysical and socio-economic components must be given equal consideration throughout all phases of EIA, including post-decision monitoring. That being said, one question that emerges from recent practice experience is: Who is responsible for **socio-economic monitoring?** Most proponents suggest that they are willing to co-operate with government and other bodies by sharing project information but maintain that socio-economic follow-up is primarily the responsibility of other parties. Mobil Oil held to this view in connection with the Hibernia offshore oil project. Similarly, in the more recent Voisey's Bay mine-mill project, the proponent took the position that the financial provisions in the impact benefits agreements to be signed with the Labrador Inuit Association and the Innu Nation were in part intended to provide those groups with the resources to carry out any socio-economic follow-up studies that they deemed necessary (Voisey's Bay Mine and Mill Environmental Assessment Panel, 1999).

MONITORING METHODS AND TECHNIQUES

Similar to impact prediction, there is no single set of monitoring techniques for all projects and components. The type of monitoring technique selected to provide data and to evaluate environmental change depends on the nature of the environmental components, purpose of the data, and particular monitoring program objectives. Thus, for both biophysical and socio-economic monitoring a variety of quantitative and qualitative techniques, or combinations thereof, are available. For example, socio-economic impacts can be monitored using annual surveys of residents' perceptions of quality of life and local police and hospital records. Selected examples of techniques for monitoring biophysical components are listed in Table 10.3. When selecting a technique for monitoring, as for any other EIA procedure, it is important to keep in mind the nature and resolution of the data required.

Table 10.3 Biophysical Monitoring Components, Parameters, and Techniques

Components	Parameters	Techniques
water quality	metals, nutrients, physical parameters	stream gauges and go-flow samplers, chemical analysis
lake biology	lake benthos	Eckman Grab stratified depth sampling and species counts
stream biology	stream benthos fish	Hester-Dendy sampler and frequency counts tagging, electro fishing, age-length keys
hydrology	water level	staff gauges, pressure transducers

KEY TERMS

ambient environmental quality monitoring
auditing
compliance monitoring
control site
cumulative effects monitoring
decision-point audit
draft EIS audit
early warning indicators
experimental monitoring
ex-post evaluation
gradient-to-background monitoring
implementation audit

inspection monitoring
life-cycle assessment
monitoring
monitoring of agreements
monitoring for knowledge
monitoring for management
performance audit
predictive technique audit
project evaluation monitoring
project impact audit
regulatory permit monitoring
socio-economic monitoring

STUDY QUESTIONS AND EXERCISES

1. What provisions exist for post-decision monitoring and auditing under your national, provincial, or state EIA system? Are these provisions mandatory or voluntary?
2. What is the value added to EIA from follow-up and monitoring activities?
3. Who should be responsible for monitoring the environment after project approval?
4. Why are socio-economic effects difficult to monitor post-project implementation? How might we address these difficulties?
5. What is the role of the public in environmental monitoring?
6. Obtain a completed project EIS from your local library or government registry, or access one on-line. Is there a monitoring component to the impact statement? What environmental aspects are included in the monitoring program? Are methods and techniques for monitoring identified? Compare your findings with those of others.

REFERENCES

Arts, J. 1998. *EIA Follow-up: On the Role of Ex-Post Evaluation in Environmental Impact Assessment*. Groningen: Geo Press.

————, P. Caldwell, and A. Morrison-Saunders. 2001. 'Environmental Impact Assessment Follow-up: Good Practice and Future Directions', *Impact Assessment and Project Appraisal* 19, 3: 175–85.

———— and S. Nooteboom. 1999. 'Environmental Impact Assessment Monitoring and Auditing', in J. Petts, ed., *Handbook of Environmental Impact Assessment*. London: Blackwell Science.

Bisset, R., and P. Tomlinson. 1988. 'Monitoring and Auditing of Impacts', in P. Wathern, ed., *Environmental Impact Assessment: Theory and Practice*. London: Unwin Hyman.

Cluff Lake Board of Inquiry. 1978. *Cluff Lake Board of Inquiry Final Report*. Regina.

Davies, M., and B. Sadler. 1990. *Post-project Analysis and the Improvement of Guidelines for Environmental Monitoring and Audit*. Report EPS 6/FA/1 prepared for the Environmental Assessment Division, Environment Canada. Ottawa.

GNWT. 2001. *Communities and Diamonds, Socio-economic Impacts on the Communities of Łutselk'e, Rae-Edzo, Rae Lakes, Wha Ti, Wekweti, Dettah, Ndilo and Yellowknife.*

Yellowknife: Annual Report of the Government of the Northwest Territories under the BHP Socio-economic Agreement.

Hegmann, G., C. Cocklin, R. Creasey, S. Dupuis, A. Kennedy, and L. Kingsley. 1999. *Cumulative Effects Assessment Practitioners Guide*. Prepared by AXYS Environmental Consulting and CEA Working Group for the Canadian Environmental Assessment Agency, Hull, Que.

LGL. 1993–7. *The Hibernia GBS Platform Construction Site Marine Environmental Effects Monitoring Program. Years 1–5*. Report prepared for the Hibernia Management and Development Company Ltd by LGL Ltd. St John's.

Minerals Management Service. 2001. *Programmatic Environmental Assessment for Grid 4. Evaluation of Kerr-McGee Oil and Gas Corporation's Development Operations Coordination Document, N-7045. Nansen Project, East Breaks, Blocks 602 and 646*. US Department of the Interior, Gulf of Mexico OCS Region.

Storey, K., and B. Noble. 2004. *Toward Increasing the Utility of Follow-up in Canadian EA: A Review of Concepts, Requirements and Experience*. Report prepared for the Canadian Environmental Assessment Agency. Hull, Que.: CEAA.

Tomlinson, P., and S. Atkinson. 1987. 'Environmental Audits: Proposed Terminology', *Environmental Monitoring and Assessment* 8, 3: 187–98.

Voisey's Bay Mine and Mill Environmental Assessment Panel. 1999. *Report on the Proposed Voisey's Bay Mine and Mill Project / Environmental Assessment Panel*. Hull, Que.: Canadian Environmental Assessment Agency.

ADVANCING PRINCIPLES AND PRACTICES IN ENVIRONMENTAL IMPACT ASSESSMENT

Cumulative Environmental Effects

DEVELOPING AN AWARENESS OF CUMULATIVE EFFECTS

The notion of cumulative environmental change is not new to EIA, and the terms 'cumulative impacts' and 'cumulative effects' actually appeared in many national EIA guidelines and legislations during the early 1970s. The US Council on Environmental Quality (1978), for example, suggested that project impacts on the environment can interact with other past, present, and reasonably foreseeable actions to generate collectively significant environmental change. It was not until the late 1980s, however, that cumulative effects started to receive attention in EIA. But even at that time it was observed that EIA had failed to consider adequately:

- the additive effects of several development projects on environmental components;
- the effects of secondary activities resulting from primary development;
- non-linear and indirect environmental responses to development;
- synergistic impacts;
- various impact interactions over time.

Only in recent years have we begun to assess systematically cumulative environmental effects in EIA practice, and considerable room remains for improvement.

Definition of Cumulative Effects
'Cumulative environmental change', 'cumulative effects', and 'cumulative impacts' are often used interchangeably. Similar to EIA, there is no universally accepted definition. Various definitions for cumulative effects have been proposed in the literature, for example:

- the accumulation of human-induced changes in VECs across space and over time that occur in an additive or interactive manner (Spaling, 1997);
- the result of individually minor but collectively significant actions taking place over a period of time (US CEQ, 1978);
- changes to the environment caused by an action in combination with other past, present, and future actions (Hegmann et al., 1999).

Sources of Cumulative Effects

The Canadian Environmental Assessment and Research Council (1988) defined cumulative effects as occurring when impacts on the biophysical or human environments take place frequently in time or densely in space to such an extent that they cannot be assimilated, or when the impacts of one activity combine with the activities of another in a synergistic manner. This suggests that a variety of different *sources* of change contribute to cumulative environmental effects. These are summarized below in Table 11.1.

Types of Cumulative Effects

While multiple types of activities and impacts can lead to cumulative environmental change, Peterson et al. (1987), Sonntag et al. (1987), and Hegmann et al. (1999) identify four broad types of cumulative effects:

1. **Linear additive effects.** Incremental additions to, or deletions from, a fixed storage where each increment or deletion has the same individual effect.

Table 11.1 Sources of Change That Contribute to Cumulative Environmental Effects

Source of change	Characteristics	Example
Space crowding	High spatial density of activities or effects	Multiple mine sites in a single watershed
Time crowding	Events frequent or repetitive in time	Forest harvesting rates exceed regeneration and reforestation
Time lags	Activities generating delayed effects	Human exposure to pesticides
Fragmentation	Changes or interruptions in patterns and cycles	Multiple forest access roads cutting across wildlife habitat
Cross-boundary movement	Effects occurring away from the initial source	Acid mine drainage moving downstream to community water supply systems
Compounding	Multiple effects from multiple sources	Heavy metals, chemical contamination, and changes in dissolved oxygen content resulting from multiple riverside industries
Indirect	Second-order effects	Decline in recreational fishery caused by decline in fish populations due to heavy metal contamination from industry
Triggers and thresholds	Sudden changes or surprises in system behaviour or system structure	The collapse of a fish stock, where persistent pressures from harvesting and environmental stress result in a sudden change in population structure

2. **Amplifying or exponential effects.** Incremental additions to, or deletions from, an apparently limitless storage or resource base where each increment or deletion has a larger effect than the one preceding.

3. **Discontinuous effects.** Incremental additions that have no apparent effect until a certain threshold is reached, at which time components change rapidly with very different types of behaviour and responses.

4. **Structural surprises.** Changes that occur due to multiple developments or activities in a defined region. These are often the least understood and most difficult to assess.

Pathways of Cumulative Effects

Cumulative environmental effects result from different combinations of actions or pathways that consist of both additive and interactive processes. Peterson et al. (1987) present a classification of functional pathways that lead to cumulative environmental effects (Figure 11.1); each pathway is identified and differentiated according to the sources of change and type of impact accumulation. An example of pathways that lead to cumulative effects is illustrated by the Cold Lake oil sands project in Alberta (Box 11.1).

Single-source perturbations. Pathway one results from the persistent effects of a single project on a particular environmental component, such as repeated changes in water temperature resulting from a reservoir development. When any single activity has multiple effects, potential interactions between them may create cumulative effects. Pathway two is characterized by a single activity, but the effects accumulate synergistically. For example, the creation of a reservoir can change water temperature, lower dissolved oxygen content, and lead to heavy metal contamination. While each of these can individually affect aquatic life, they can also accumulate in such a way that the toxicity of certain contaminants is multiplied because of high water temperatures and low dissolved oxygen content (Bonnell, 1997).

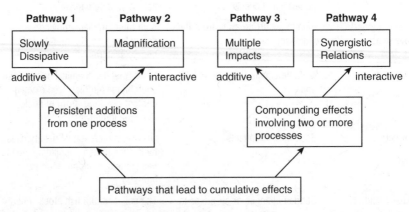

Figure 11.1 Pathways Leading to Cumulative Effects

Source: Based on Peterson et al. (1987).

Box 11.1   Cold Lake Oil Sands Project Cumulative Effects Pathways

The Cold Lake oil sands project is a heavy oil facility in northern Alberta. Approximately 2,500 wells are currently operating in the region. The Cold Lake facility is currently the second largest producer of oil in Canada. In 2003 the Cold Lake operations accounted for 10 per cent of Canada's crude oil production. Oil deposits are located in sand deposits approximately 400 metres below the surface and are extracted by a steam recovery process that injects high-pressure steam into the reservoir to separate the sand and oil. Wells are drilled and steam injected via clusters of vertical and directional drilled wells, organized onto large surface pads. In 1997 the proponent, Imperial Oil Resources Limited, proposed to expand its operation in the Cold Lake area with the development of a central plant and additional production wells. A total of 35 impact models were contained in the EIA to assess the cumulative effects of the project on surface water quality, including the additive effects of roads and facilities (well pads) on sediment and contaminant levels in nearby water bodies.

Cumulative impact statement:
- Operation and maintenance of roads and facilities will result in the generation of sediment and transport of contaminants to receiving waters.

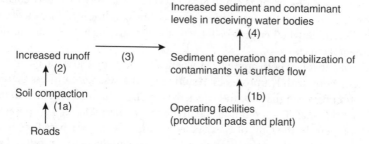

Pathways:

1a. The operation and maintenance of roads will lead to compaction of the roadbed.
1b. Operation and maintenance of pads and plant facilities will result in the generation of sediment and mobilization of contaminants via overland flow from these facilities.
2. Compaction will cause an increase in surface runoff from the road.
3. Increased runoff from roads will result in erosion of exposed soils, resulting in an increase in sediment generation and transport. Soluble contaminants from the road and the roadbed will be transported along with the sediment.
4. Increased sediment and contaminant transport will result in higher levels of these parameters in receiving waters, which will result in a decline in surface water quality.

Sources: Based on Hegmann et al. (1999); Imperial Oil Resources Ltd (1997).

Accumulation of effects from two or more projects. Pathway three occurs when the environmental effects of multiple actions accumulate in an additive manner, as would be the case with the development of multiple reservoirs in a river basin. Although no interaction occurs between the effects of individual projects, they collectively result in significant impacts on aquatic resources. Pathway four occurs when these multiple effects do interact in a synergistic manner. For example, each project may alter water temperatures, change dissolved oxygen content, and introduce heavy metals, thereby contaminating aquatic life, but the impacts from the interaction of these effects across all projects would be greater than the sum of the individual project impacts (Bonnell, 1997).

CUMULATIVE EFFECTS ASSESSMENT

It has been said by many that cumulative effects assessment (CEA) is simply EIA done right. In practice, however, CEA is often done after the initial identification of effects have been completed in a project EIA. Dube (2003), for example, distinguishes between two broad approaches to cumulative effects assessment—effects-based and stressor-based.

Effects-based CEA concerns measuring actual environmental responses to stressors, including unidentified stressors and multiple stressor interactions. Effects are then compared to some reference condition or control point to determine an actual measure of cumulative change, irrespective of the number and nature of the impacts causing that change.

Stressor-based CEA predicts cumulative effects associated with a particular agent of change. This involves the description of a baseline condition and predictive modelling to determine whether project-related stressors are likely to create cumulative environmental effects.

Ideally, CEA involves both approaches where all aspects of a stressor-based CEA are done concurrently with EIA processes, resulting in an assessment process that makes no distinction between the two, and effects-based CEA is then directly tied to project and broader regional effects monitoring. In the case of Kerr-McGee's proposal for oil well development in the Gulf of Mexico, for example, a stressor-based approach to CEA was adopted with the expectation that the assessment document produced would serve as a reference point for effects-based CEA for future activities and development within the Gulf region (Box 11.2).

There is no universally accepted framework for assessing cumulative environmental effects, but, similar to EIA, a generic process for CEA can be identified (Table 11.2).

Box 11.2 Stressor- and Effects-based CEA in the Gulf of Mexico, US Minerals Management Service

The US Minerals Management Service (MMS), a bureau of the Department of Interior, is the responsible federal government agency for managing the country's natural gas, oil, and other mineral resources on the outer continental shelf. The MMS's Offshore Minerals Management Program currently has in place a comprehensive strategy for post-lease National Environmental Policy Act (NEPA) compliance in deepwater areas in the central and western areas of the Gulf of Mexico. Under this strategy, the MMS prepares a **programmatic environmental assessment** to address proposed development projects within the Gulf

Continued

region. Once the programmatic impact assessment is completed, its function is to serve as a reference document for subsequent environmental analyses focused on specific projects and environmental components (MMS, 2001).

In 2001 the US MMS Gulf of Mexico OCS Region prepared a programmatic EA of Kerr-McGee Oil and Gas Corporation's proposal to complete and produce in the Gulf region a total of 12 wells drilled under previously approved exploration plans. Kerr-McGee Corporation is a global energy and inorganic chemical producer based in Oklahoma City. All of the proposed wells, with the exception of three sub-wells, would share a common surface location, but oil and gas from nearby projects would be commingled with the proposed initiative through a series of transmission pipelines. The purpose of the assessment was to determine the specific and cumulative impacts associated with proposed oil and gas development and production within the East Breaks administrative region of the Gulf of Mexico. Aside from approval with existing and/or added mitigation, the only alternative to the proposal was that of 'no approval.'

The MMS previously addressed the cumulative effects of activities on the outer continental shelf and non-outer continental shelf for the Gulf Coast region for the years 1996 through 2036 as part of the NEPA process for proposed multi-sale/lease activities. The proposed Kerr-McGee project was assessed in relation to current and existing projects, activities, and cumulative effects within the region. In particular, cumulative effects for the region were considered for water quality, air quality, sensitive coastal environments, deepwater benthic communities, marine mammals, birds, fish resources, economic and demographic conditions, commercial fisheries, and archaeological resources. The MMS concluded that the proposal would have no significant cumulative impact on the regional resources, and the preparation of a detailed environmental impact statement was not required.

Source: MMS (2001).

Table 11.2 Generic Assessment Framework for CEA

Basic EIA steps	CEA tasks
1. Scoping	• Identify regional issues of concern • Select appropriate regional VECS • Identify VEC objectives and indicators • Identify spatial and temporal boundaries • Identify other actions that may affect the same VECS • Identify potential impacts due to actions and possible effects
2. Analysis of effects	• Complete regional baseline data collection • Assess effects of proposed action on VECS • Assess effects of all combined actions in the region on VECS
3. Impact management	• Identify impact management and mitigation strategies
4. Determine significance	• Evaluate the significance of residual effects • Compare results against thresholds and objectives
5. Monitor	• Identify VEC indicators for effects monitoring • Develop regional effects monitoring program

Source: Based on Hegmann et al. (1999).

Perhaps the most significant areas of deviation from conventional project EIA involve establishing the spatial and temporal boundaries for assessment and the application of assessment methods. Cumulative effects over a large spatial scale and over considerably longer time frames, particularly those 'Pathway 4' effects identified in Figure 11.1, are extremely difficult to assess.

Spatial Boundaries for CEA

> If we pick the right boundaries, we have a better chance of addressing what's going on in the proper scale. (Beanlands and Duinker, 1983: 49)

Determining the spatial boundaries for a CEA is critical to its success in effectively managing the cumulative impacts associated with development projects. Boundaries in CEA delimit the spatial extent of the assessment and thus the environments and VECs that are considered. It is generally acknowledged that in order to assess cumulative effects effectively there is a need to extend the spatial boundaries of the assessment well beyond the project site. However, if large boundaries are defined, only a superficial assessment may be possible and uncertainty will increase. Moreover, the incremental additions of a single project may seem less and less significant—only a small drop in a large bucket. However, if the boundaries are small a more detailed examination may be feasible but an understanding of the broad context may be sacrificed. In addition, the incremental impacts of a single project may be exaggerated— a large drop in only a small bucket (Figure 11.2). This, of course, depends on the nature of the project and the impacts. For example, by selecting a relatively small spatial boundary the impacts of emissions from a proposed smelting operation might seem quite insignificant, especially if emission stacks are relatively high. The choice of spatial boundaries for CEA, then, can be to the proponent's advantage or disadvantage.

Establishing the appropriate boundaries for CEA requires consideration of three types of scale:

Spatial scale refers to the actual geographic extent of the assessment and is typically based on one or both of natural boundaries, such as watersheds, or administrative boundaries, such as townships or landownership. This is usually the most common interpretation of 'scale' in CEA; however, it is certainly not the most functional with regard to the actual analysis of cumulative effects.

Analysis scale is used to examine VECs and impacts across space, and is represented by such ideas as data resolution, detail, and granularity. In the Cold Lake oil sands project discussed in Box 11.1, for example, the geographic boundaries for wildlife and vegetation were restricted to local township areas based on the availability of historical and current information on vegetation composition and wildlife habitat, as well as the extent of available aerial photo coverage. Spatial boundaries were determined based on data availability and desired analysis scale.

Phenomenon scale is perhaps the most important type of scale in CEA, as it refers to the spatial units within which various processes operate or function. Thus, in any

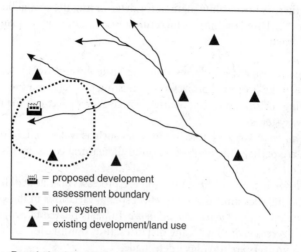

Restrictive bounding may result in the project's additive effects on water quality being perceived as quite *significant* when considered together with the surrounding development and land-use activities.

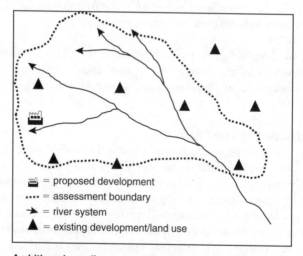

Ambitious bounding may result in the project's additive effects on water quality being perceived as quite *insignificant* when considered in light of the total effects of all development and land-use activities on the watershed.

Figure 11.2 Restrictive and Ambitious Spatial Bounding

single assessment different spatial boundaries may be appropriate for different cumulative effects and for different VECs. The boundaries selected for cumulative effects on air quality might be quite different from those chosen to assess cumulative effects on soil quality or sedentary versus migratory wildlife.

While there is no absolute method for determining the spatial boundaries for any particular CEA, based on the CEA literature a number of guiding principles emerge to assist the practitioner.

- *Adequate scope.* Boundaries must be large enough to include relationships between the proposed project, other existing projects, and the VECs. This means crossing jurisdictional boundaries if necessary to account for interconnections across systems.
- *Natural boundaries.* Natural boundaries such as watersheds, airsheds, or ecosystems are perhaps the best reflection of the natural components of a system and should be respected.
- VEC *differentiation.* Different VECs and VEC processes operate at different spatial scales and boundaries must therefore reflect spatial variations in the VECs considered.
- *Maximum zones of detectable influence.* Impacts related to project activities typically decrease with increasing distances, thus boundaries should be established where impacts are no longer detectable.
- *Multi-scaled approach.* Multiple spatial scales, such as local and regional boundaries, should be assessed to allow for a more in-depth understanding of the scales at which VEC processes and impacts operate.
- *Flexibility.* CEA boundaries must be flexible to accommodate changing natural and human-induced environmental conditions.

Generally speaking, however, approaches to cumulative effects assessment seem to be most successful with clearly bounded systems, such as lakes and watersheds, than with more open systems, such as estuaries, marine waters, and terrestrial systems (CEARC, 1988).

Temporal Boundaries for CEA

At the heart of cumulative effects assessment is the consideration of the influence of other past, proposed, or likely future activities. Establishing temporal boundaries for CEA requires asking 'how far back in time' and 'how far into the future' should cumulative environmental change resulting from the proposed and other actions and activities be considered in the assessment. The extent of temporal boundaries depends on the amount of information desired, the amount of information available, and what the assessment is trying to accomplish. Examining past conditions may be as simple as examining land-use maps, and in certain cases it may be feasible to incorporate 50 years of historical data if deemed necessary.

The Canadian Environmental Assessment Agency's *Cumulative Effects Assessment Practitioners' Guide* (Hegmann et al., 1999) outlines several options for establishing how far into the past a CEA should extend. The first two options have limited historical perspective and are based on the temporal characteristics of the proposed project itself:

- temporal bounds established based only on the existing environmental conditions; or
- when impacts associated with the proposed action first occurred.

Other options are based on more historical perspectives of land use and conditions of environmental change, and include:

- the time at which a certain land-use designation was made (for example, the establishment of a park or the lease of land for development);
- the point in time at which effects similar to those of concern first occurred; or
- a past point in time representative of desired environmental conditions or pre-disturbance conditions, especially if the assessment includes determining to what degree later actions have affected the environment.

CEA boundaries for *future conditions* are often based on:

- the end of operational life of the proposed project;
- the point of project abandonment and site reclamation; or
- a time when VECs are likely to be recovered, considering natural variations, to their to pre-disturbance conditions.

Identifying potential future actions and activities to include in CEA can be much more uncertain. Hegmann et al. (1999) thus characterize future actions of three types:

1. *Certain actions.* The action will proceed or there is a high probability the action will proceed. This includes actions or projects already approved or submitted for approval, or that have been proposed by the proponent.
2. *Reasonably foreseeable actions.* The action may proceed, but there is some uncertainty about this conclusion. This might include projects under review for which approval is likely to be conditional, activities identified in an approved or proposed development plan, or induced activities that may occur should the project proposed be approved.
3. *Hypothetical actions.* There is considerable uncertainty whether the action will ever proceed. Such actions or activities include those discussed on only a conceptual basis or those speculated based on current information. **Scenario analysis** is a common tool used to identify such hypothetical actions.

These actions lie on a continuum from most likely to least likely to occur. For each assessment, the practitioner or the regulatory agency will have to decide how far into the future the assessment should reach. Often, a major criterion is whether the future action or actions are likely to affect the same VECs as the proposal under consideration. While practical, this criterion may detract from those projects, creating 'nibbling' effects that, while they may not directly affect the same VECs, contribute to overall decline in environmental quality.

Methods for CEA
Many of the methods used in EIA are equally applicable to addressing cumulative environmental effects (Table 11.3). The difficulties inherent in each method still exist, however, and in fact seem to be exacerbated by the complexity associated with the

Table 11.3 Selected Methods That Support Cumulative Effects Assessment

Method	Characteristics
Geographic information systems	Provide spatial analysis of cumulative effects through digital mapping. Useful for mapping sources of cumulative change and effects, and areas of overlapping project impacts. Limited application for analysis of pathways, and restricted by data availability.
Network analysis	A qualitative network based on feedback relationships. Useful for identification of second-, third-, and higher-order effects pathways and relationships, but largely untested in CEA practice.
Interactive matrices	Total cumulative impact is assumed to be the sum of project-specific effects derived from overlain or otherwise manipulated impact identification and assessment matrices. Allows the consideration of cumulative effects of multiple sources, but effects are not differentiated by type. Values rely heavily on expert judgement.
Ecological modelling	Computer-based, dynamic modelling of ecosystem components and interactions. Theoretically sound, can accommodate a large volume of data, and easily provides opportunity for scenario analysis of future conditions. Application, however, is dependent on data availability, model validation, and verification of relationships between VECs, and is limited to systems for which behaviour is well understood.
Expert weighting and scoring	Assignment of impact scores and impact values to affected VECs to derive a total cumulative impact score. The approach is not limited by data availability, and provides a simple, sound basis on which to evaluate the additive impacts of proposed developments across VECs. However, the impact scores and weights are only as informative as the individual who assigns them.

Source: Based on Canter (1999).

nature of cumulative effects. For example, the subjectivity of weighting values in matrix techniques becomes even more problematic when multiple projects are being considered, and the coarseness of data and its analysis often increase as the spatial scale of the assessment increases. Two methods previously discussed in Chapter 3 are reiterated here within the context of simple additive CEA.

Interaction matrices. Interaction matrices can be used to identify the effects of multiple projects. This requires the subjective assignment of magnitude and significance values to the impacts expected to result from each individual project or from different stages of the same project, and 'summing up' the overall cumulative impacts. In Figure 11.3 a series of impact assessment matrices are shown for both the cumulative effects of a single activity and the cumulative effects of multiple activities on selected VECs. Each matrix is first evaluated individually, based on its magnitude and signifi-

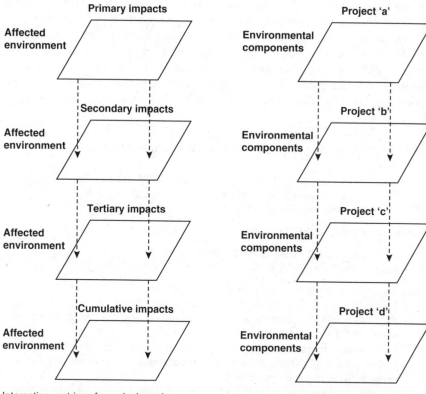

Interaction matrices for a single project generate a cumulative matrix depicting the total primary, secondary, and tertiary effects on particular environmental components. For example, clearing of land may increase erosion (primary impact), which in turn increases runoff (secondary impact), leading to increased stream turbidity (tertiary impact). Similar project activities may similarly generate effects on water quality, contributing to overall cumulative impact.

In a similar fashion, matrices can be overlain to sum the impacts of multiple projects on various environmental components. Although not depicted here, each environmental component can be weighted according to its relative importance, with impact magnitude and significance values incorporated into each project matrix. This approach is useful for proving an overall view of impacts associated with activities in a particular geographic area.

Figure 11.3 Matrix Interactions for Multiple Components of the Same Project and Multiple Projects

Source: Based on Dixon and Montz (1995).

cance, and is then overlain or summed with each other matrix to generate an overall perspective of cumulative environmental effects.

Weighting methods. A common approach to CEA for simple additive impacts involves weighting methods and techniques. The Cluster Impact Assessment Procedure (CIAP), for example, was developed by the US Federal Energy Regulatory Commission for the CEA of small-scale hydroelectric power projects. At the heart of the CIAP is the calculation of a 'total impact' score for each of the affected VECs based on individual project impacts and related interaction impacts. Numerical impact ratings are assigned as impact scores and the relative importance of the components is considered (Canter, 1999). A similar approach, using a Delphi process, was adopted by Bonnell (1997) to assess the cumulative environmental effects of multiple existing and proposed small-scale hydroelectric development projects in Newfoundland. An example of weighting impacts and VECs for simple additive effects is depicted in Box 11.3.

Box 11.3 Weighting Impacts and VECs for Simple Additive CEA

Projects: small-scale hydroelectric project '**A**'
small-scale hydroelectric project '**B**'
small-scale hydroelectric project '**C**'

VECs: forest resources, fish resources, caribou

Impact of project 'n' on VEC 'i' $\quad I_nVEC_i = P_i(M_i \times SE_i \times TD_i) \times CS_i$

$P_i =$ the probability that VEC 'i' will be affected by project 'n'
- expressed as % (i.e., likelihood)

$M_i =$ the magnitude of the effect of a project 'n' on VEC 'i'
- 1 = negligible: a change to the VEC that is indistinguishable from natural variation
- 2 = minor: a reversible change to the VEC's normal baseline condition; the fundamental integrity of the VEC is not affected
- 3 = moderate: a reversible change to the VEC's normal or baseline condition, with a medium probability of second-order effects on other environmental components; the fundamental integrity of the VEC is not threatened
- 4 = major: an irreversible change to the VEC's normal or baseline conditions, with a high probability of second-order effects on other environmental components; the fundamental integrity of the VEC is threatened

$SE_i =$ the spatial extent of the effect of project 'n' on VEC 'i'
- 1 = site-specific: effect will be confined to the project development area
- 2 = local: effect will be confined to the project area and immediate environment
- 3 = regional: effect will occur within and beyond the development area and immediate environment, affecting a defined territory surrounding the development
- 4 = ecosystem: effect will occur throughout the ecosystem

$TD_i =$ the temporal duration of the effect of a project 'n' on VEC 'i'.
- 1 = short-term: effect may persist less than two years from the onset of disturbance
- 2 = medium-term: effect may persist from two to less than five years from the onset of disturbance
- 3 = long-term: effect may persist from five to less than 10 years from the onset of disturbance
- 4 = prolonged: effect may persist 10 years of more from the onset of disturbance

$CS_i =$ the current state of VEC 'i' in the vicinity of project 'n'.
- 1 = resilient: VEC is quite resilient to impact, due to its natural conditions and/or lack of other adjacent human activities
- 2 = low sensitivity: VEC has a low susceptibility to impact, due to its natural condition and/or the impacts of other adjacent human activities

Continued

- 3 = medium sensitivity: VEC is moderately susceptible to impact, due to its natural conditions and/or the impacts of other adjacent human activities
- 4 = high sensitivity: VEC is highly susceptible to impact, due to its natural condition and/or the impacts of other adjacent human activities

VEC Imp 'i' = the relative importance of VEC 'i' compared to all other VECs with regard to (for example):
   a) socio-economic dependency
   b) human health
   c) ecological functioning

Step 1: Determine relative importance (significance) of affected VECs based on:

a) socio-economic dependency

Using Saaty's paired comparison methods, discussed in Chapter 7, the following matrices present a hypothetical example of VEC weighting. For example, in terms of socio-economic dependency, 'forest resources' (row 1) are considered slightly less important than 'fish resources' (column 2) ('1/3'), and significantly less important than 'caribou' (column 3) ('1/5').

	Forest resources	Fish resources	Caribou
Forest resources	1	1/3	1/5
Fish resources	3	1	1/3
Caribou	5	3	1

Divide each cell entry of the paired comparison matrix by the sum of its corresponding column to normalize the matrix:

	Forest resources	Fish resources	Caribou
Forest resources	0.11	0.08	0.38
Fish resources	0.33	0.23	0.63
Caribou	0.55	0.69	1.88

Determine the priority vector of the matrix by averaging the row entries in the normalized assessment matrix.

- Forest resources = 0.19; Fish resources = 0.40; Caribou = 1.04

Multiply each column of the initial assessment matrix by its relative priority, and sum the results:

$$0.19 \begin{pmatrix} 1 \\ 3 \\ 5 \end{pmatrix} + 0.40 \begin{pmatrix} 1/3 \\ 1 \\ 3 \end{pmatrix} + 1.04 \begin{pmatrix} 1/5 \\ 1/3 \\ 1 \end{pmatrix} = \begin{pmatrix} 0.53 \\ 1.32 \\ 3.19 \end{pmatrix}$$

Divide the results by the original priorities to determine the VEC importance based on socio-economic dependency:

- Forest resources = 0.53 / 0.19 = 2.70
- Fish resources = 1.32 / 0.40 = 3.30
- Caribou = 3.19 / 1.04 = 3.08

Continued

Follow the same procedure for **b) human health** and **c) ecological functioning** to determine overall VEC importance. For example:

	Socio-economic dependency	Human health	Ecological functioning	Total VEC importance*
Forest resources	2.70	1.83	1.45	5.98
Fish resources	3.30	2.15	1.24	6.69
Caribou	3.08	3.24	1.89	8.21

*The assumption is that 'socio-economic dependency', 'human health', and 'ecological functioning' are equally important. Otherwise, these factors must be weighted as well and the weight multiplied by each of the respective VEC weights before adding the 'total VEC importance'.

2. Determine the impact of each project on each VEC by '$I_nVEC_i = P_i(M_i \times SE_i \times TD_i) \times CSi$'

For example, for project 'A'

	P_i	M_i	SE_i	TD_i	CSi	$I'A'VEC$
Forest resources	0.45	3	1	2	3	8.1
Fish resources	0.10	4	3	3	4	14.4
Caribou	0.78	2	2	2	3	18.72

Follow the same procedure for projects 'B' ($I'B'VEC$) and 'C' ($I'C'VEC$) and transfer all values to the total cumulative effects score table shown below.

3. Determine the weighted and cumulative impacts

Project 'a' impact on VEC 'i' $\times$ Imp VEC 'i'

	$I'A'VEC \times$ total VEC importance	$I'B'VEC \times$ total VEC importance	$I'C'VEC \times$ total VEC importance	Cumulative (additive) impact score
Forest resources	(8.1)(5.98) = 48.44	(2.21) (6.12) = 13.52	(5.21)(6.12) = 31.89	Σ = 93.85
Fish resources	(14.4)(6.69) = 96.34	(3.24)(6.57) = 21.29	(4.21)(6.57) = 27.66	Σ = 145.29
Caribou	(18.72)(8.21) = 153.69	(7.12)(8.46) = 60.23	(3.12)(8.46) = 26.40	Σ = 240.32
Cumulative impact of project 'n' on all VECs =	Σ = 298.47	Σ = 95.04	Σ = 85.95	Overall cumulative impact of all projects on all VECs Σ = 479.46

This approach allows quick insight as to the individual contributions of each project to the overall cumulative effects. Project 'A', for example, is contributing proportionately more to overall cumulative impacts when compared to projects 'B' and 'C'. Moreover, it is generating the most significant effects on caribou, the VEC deemed to be of greatest overall importance to socio-economic dependency, health, and ecological functioning. While this approach is highly subjective and perhaps no more accurate than the individual(s) assigning the values, it does allow for a systematic examination of the additive effects of individual projects on VECs and establishes a baseline from which effects-based CEA can be undertaken. The approach is also useful in situations where baseline data are not available, or to analyze the judgements of experts, publics, or other project interests.

Sources: Based on Irving and Bain (1993); Bonnell (1997); Canter (1999); Noble (2002).

TOWARDS A REGIONAL APPROACH TO CEA IN CANADA

Section 16 (1)(a) of the Canadian Environmental Assessment Act requires that every project assessment consider the environmental effects of the project, including any cumulative environmental effects that are likely to result from the project in combination with other projects or activities that have been or will be carried out. Traditionally, cumulative effects have been addressed through project-based EIA. The challenges and limitations associated with project-based approaches, however, are now widely documented (Cooper, 2003; Creasey, 2002). This is not to say that project-based CEAs are not useful; rather, something more is needed to address and manage cumulative environmental change in an effective manner. There is now general agreement that CEA should go beyond the evaluation of site-specific direct and indirect project impacts to address broader regional environmental impacts and concerns. Amendments to the Act in 2003, for example, gave particular attention to the value of regional studies in improving the systematic consideration of effects.

Recognizing that each additional project represents a high marginal cost to the regional environment, regional approaches to CEA attempt to expand the spatial boundaries to address overall impacts on the region due to multiple perturbations over time and space. **Regional CEA** focuses on a wider range of impacts resulting from multiple project developments and environmental component interactions within a spatially defined area. The underlying notion is that cumulative environmental change is the product of multiple, interacting development actions and that the multiplicity of development decisions in a particular region, while often individually insignificant, cumulatively lead to significant environmental change (Smit and Spaling, 1995). Cocklin et al. (1992) identify three main objectives in pursuing a regional approach to cumulative effects assessment:

- to develop a broader understanding of the current state of the environment vis-à-vis cumulative change processes;
- to identify, insofar as possible, the extent to which cumulative effects in the past have conditioned the existing environment;
- to consider priorities for future environmental management with respect to general policy objectives and with regard to potential development options.

Considerable challenges to regional CEA remain (Box 11.4). Part of the problem is that project proponents cannot be held responsible for the actions of other developers, nor can they be expected to acquire information on other development projects in the region—especially when proponents are not always willing to share project information. Moreover, CEA is not explicitly objectives-oriented, and it is not designed to be part of a broader regional environmental management policy or planning framework. The traditional approach to regional CEA has been to address the symptoms or effects of development, including the effects of multiple project developments in a particular administrative region after development is proposed or initiated, rather to address higher-order strategic initiatives and objectives in order to identify the sources of change before irreversible decisions are taken. A more proactive and objectives-led approach is required to address sources of regional cumulative environmental effects than what is currently achieved through conventional EIA-driven frameworks.

Box 11.4 Need for Better-Practice Regional CEA: Terra Nova Offshore Oil Project

With an estimated field reserve of over 400 million barrels, the Terra Nova oil field, discovered in 1984, is located approximately 350 km east-southeast of St John's, Newfoundland, and 35 km southeast of the Hibernia oil field. In 1996, Petro-Canada, primary operator and 34 per cent shareholder, submitted a development application for approval. The proponents commenced public hearings in April 1997 and in August 1997, after taking into consideration the proponent's impact statement, submissions by government agencies, non-governmental groups, and members of the public, a panel recommended approval of the project. Production of the field began in January 2002.

One contentious issue that emerged during public hearings was the scope of CEA and the responsibility of the proponent. Public submissions to the review process emphasized the cumulative and synergistic concerns in the offshore area, including fishery depletions, oceanographic changes, seabird hunting, climate change, and transportation. Potential future petroleum developments on the Grand Banks, including specific project developments within the region and broader plans to establish a larger offshore industry, were also identified as matters that should be considered in evaluating the project's cumulative impacts. The proponent, in contrast, argued that cumulative impacts should be considered for the Terra Nova project only in terms of specific, planned petroleum projects on the Grand Banks. The panel agreed, suggesting that it is not possible to hold the proponents responsible for future developments beyond their control that may interact with the Terra Nova project to produce cumulative environmental effects. It was suggested during the public review process that an additional comprehensive environmental assessment of all the proposed and potential offshore developments was necessary to include, among other things, the rapidly changing fishery conditions in light of petroleum developments.

KEY TERMS

ambitious bounding
amplifying effects
analysis scale
discontinuous effects
effects based CEA
linear additive effects
phenomenon scale

programmatic environmental assessment
regional CEA
restrictive bounding
scenario analysis
spatial scale
stressor-based CEA
structural surprises

STUDY QUESTIONS AND EXERCISES

1. Do provisions exist under your national, provincial, or state EIA system for cumulative effects assessment?
2. Using the example of multiple reservoir developments in a single watershed, sketch a diagram similar to Figure 11.1 and identify and classify the different types of cumulative impacts that might result. State the impact 'pathways' as illustrated by the example in Box 11.1.

3. What is the difference between effects-based and stressor-based approaches to cumulative effects assessment?

4. Using an example, explain how a proponent might use spatial bounding to its advantage. Given this, should the proponent be solely responsible for determining the spatial boundaries for cumulative effects assessment?

5. It has been said that cumulative effects assessment is simply EIA done right. Using Box 1.4, 'Generic EIA Process', and Box 11.3, 'Generic Assessment Framework for CEA', construct a new framework that integrates the principles of both EIA and CEA. You might want to start by first listing the basic EIA steps, and then for each step attempt to integrate CEA tasks.

6. Cumulative effects often result from multiple and often unrelated project developments in a single region. Should regional cumulative effects assessment be the responsibility of the project proponent?

7. When a project creates environmental damage it is often the responsibility of the proponent to rectify or compensate for such damage. Assume a region where there exist multiple projects and activities, including oil and gas, forestry, highways, recreation, and hydroelectric developments. Individually, each project was approved for development based on the fact that it would not generate significant environmental effects. Cumulatively, however, all of these activities are contributing to overall environmental decline.

   a) Who should be responsible for managing overall cumulative environmental change resulting from the many, unrelated project developments and activities?

   b) How does one determine how much each development or activity is contributing to cumulative change?

   c) Given that each project is 'individually insignificant' but cumulatively damaging, should an additional development be permitted for the region if it, too, is determined to be individually insignificant? What are the implications of such a decision with regard to equity versus environmental protection?

8. Assume that three industrial developments are proposed for a previously undeveloped region adjacent to a river. Each project is similar in size and each will require clearing of a forested site. Several VECs and their overall importance weightings have been identified for the region, including wildlife (2.21), aquatic resources (3.41), water quality (5.73), and air quality (4.68). Using the equation 'InVECi = Pi(Mi × SEi × TDi) × Csi', construct a hypothetical cumulative impact assessment matrix for each project. Assign hypothetical project impact scores and calculate the following:

   a) the cumulative impact of each individual project across all VECs;

   b) the cumulative impact of all projects on each VEC;

   c) the cumulative impact of all projects across all VECs;

   Discuss the advantages and limitations of the above approach to cumulative effects assessment.

REFERENCES

Beanlands, G.E., and P.N. Duinker. 1983. 'Lessons from a Decade of Offshore Environmental Impact Assessment', *Ocean Management* 9, 3/4: 157–75.

Bonnell, S. 1997. 'The Cumulative Effects of Proposed Small-scale Hydroelectric Developments in Newfoundland, Canada', MA thesis, Memorial University of Newfoundland.

Canadian Environmental Assessment Research Council (CEARC). 1988. *The Assessment of Cumulative Effects: A Research Prospectus*. Ottawa: Supply and Services.

Canter, L. 1999. 'Cumulative Effects Assessment', in J. Petts, ed., *Handbook of Environmental Impact Assessment*, vol. 1, *Environmental Impact Assessment: Process, Methods and Potential*. London: Blackwell Science.

Cocklin, C., S. Parker, and J. Hay. 1992. 'Notes on Cumulative Environmental Change I: Concepts and Issues', *Journal of Environmental Management* 35: 51–67.

Cooper, L. 2003. *Draft Guidance on Cumulative Effects Assessment of Plans*. EPMG Occasional Paper 03/LMC/CEA. London: Imperial College.

Creasy, R. 2002. 'Moving from Project-based Cumulative Effects Assessment to Regional Environmental Management', in A.J. Kennedy, ed., *Cumulative Environmental Effects Management: Tools and Approaches*. Calgary: Alberta Society of Professional Biologists.

Dixon, J., and B.E. Montz. 1995. 'From Concept to Practice: Implementing Cumulative Impact Assessment in New Zealand', *Environmental Management* 19, 1: 445–56.

Dube, M. 2003. 'Cumulative Effects Assessment in Canada: A Regional Framework for Aquatic Ecosystems', *Environmental Impact Assessment Review* 23: 723–45.

Hegmann, G., C. Cocklin, R. Creasey, S. Dupuis, A. Kennedy, and L. Kingsley. 1999. *Cumulative Effects Assessment Practitioners' Guide*. Prepared by AXYS Environmental Consulting and CEA Working Group for the Canadian Environmental Assessment Agency, Hull, Que.

Imperial Oil Resources Limited. 1997. *Cold Lake Expansion Project Environmental Impact Statement*. Calgary: Imperial Oil Resources Limited.

Irving, J.S., and M.B. Bain. 1993. 'Assessing Cumulative Impact on Fish and Wildlife in the Salmon River Basin, Idaho', in S.G. Hilderbrand and J.B. Cannon, eds, *Environmental Analysis: The NEPA Experience*. Boca Raton, Fla: Lewis Publishers.

Minerals Management Service. 2001. *Programmatic Environmental Assessment for Grid 4. Evaluation of Kerr-McGee Oil and Gas Corporation's Development Operations Coordination Document, N-7045. Nansen Project, East Breaks, Blocks 602 and 646.* US Department of the Interior, Gulf of Mexico OCS Region.

Noble, B.F. 2002. 'Strategic Environmental Assessment of Canadian Energy Policy', *Impact Assessment and Project Appraisal* 20, 3: 177–88.

Peterson, E., et al. 1987. *Cumulative Effects Assessment in Canada*. Ottawa: Supply and Services Canada.

Smit, B., and H. Spaling. 1995. 'Methods for Cumulative Effects Assessment', *Environmental Impact Assessment Review* 15: 81–106.

Sonntag, N., et al. 1987. *Cumulative Effects Assessment: A Context for Further Research and Development*. Ottawa: Supply and Services Canada.

Spaling, H. 1997. 'Cumulative Impacts and EIA: Concepts and Approaches', *EIA Newsletter* (University of Manchester) vol. 14. Available at: <www.art.man.ac.uk/EIA/publications/index.htm>.

US Council on Environmental Quality (CEQ). 1978. *Considering Cumulative Effects Under the National Environmental Policy Act*. Washington: Council on Environmental Quality, Executive Office of the President.

Strategic Environmental Assessment

THE BASIS OF STRATEGIC ASSESSMENT

Strategic environmental assessment (SEA) has become one of the most widely discussed issues in the field of contemporary environmental assessment. It is difficult to cover all aspects of SEA in a single chapter, as SEA itself deserves its own separate manual. Attention here is thus limited to selected SEA principles and practices: basic SEA characteristics, the international status of strategic environmental assessment, and the illustration of a practical framework for application. Sheate et al. (2001) provide a comprehensive discussion of SEA, including country status reports covering European Union countries, in *SEA and Integration of the Environment into Strategic Decision-making*. This report is available on the European Commission SEA Web site at <http://europa.eu.int/comm/environment/eia/sea-support.htm>.

SEA is based on the notion that many of the decisions that affect the environment are made long before project developments are proposed. Moreover, such decisions often affect the nature and type of project proposals. For example, decisions concerning a regional land-use management plan are likely to shape the future development of the region and the specific types of project activities undertaken. **Strategic environmental assessment**, while variously defined, broadly refers to the environmental assessment of **policies, plans, programs** (PPPs), and their alternatives. In essence, SEA extends EIA upstream—but at the same time it adopts a different set of principles than those used in project-based assessment (Box 12.1).

SYSTEMS OF SEA

While the origins of SEA are often tied to the US NEPA of 1969, it was not until after a number of high-profile international developments, namely the World Bank's (1999) recommendation for environmental assessment of policy, the report of the World Commission on Environment and Development, *Our Common Future* (1987), and the United Nations' Earth Summit held in 1992 in Rio de Janeiro, that SEA gained international recognition. It is only within the past decade, however, that SEA has been clearly evidenced in practice.

Currently, SEA is far less advanced than EIA and only a few nations have formal provisions for SEA systems. Where provisions do exist, they vary considerably

---

Box 12.1 Basic Principles of Strategic Environmental Assessment

Strategically focused	• emphasis is placed on identifying a strategy for action • determines objectives and the means to achieve them
Future-oriented	• forward, long-term planning perspective • identifies desired ends and alternative futures
Focused on alternatives	• assessment of 'alternative options' and strategies • examines alternatives to meet a need or to address a problem • examines alternatives to a proposed or existing policy, plan, or program
Objectives-led	• examines particular goals and objectives to be accomplished • assessment is set within the context of a broader vision
Proactive	• attempts to avoid, eliminate, and minimize potentially negative actions • attempts to enhance and create potentially positive actions • creates and examines alternatives to identify the preferred course of action • commences early, before irreversible decisions are made
Integrated	• incorporates multiple objectives, criteria, and sources of knowledge • ideally, an integral part of policy, plan, and program formulation
Broad focus	• not project specific • more broad-brush than project-level assessments • scope broadens as assessment moves from program, to plan, to policy level
Tiered	• set within the context of previous and subsequent decision outcomes and objectives • sets the stage for subsequent assessment and decision-making processes

Source: Based on Noble (2000).

(Box 12.2). In the US, for example, provisions for SEA fall under the National Environmental Policy Act of 1969, where SEA is broadly interpreted to be *programmatic environmental assessment* or area-wide EIA. Hundreds of programmatic assessments are completed in the US each year, which essentially involve direct application of EIA to plans and programs.

In the UK, SEA was traditionally carried out through a less formalized policy and plan environmental appraisal process. Formal requirements for SEA were adopted in the UK in 2004 under European SEA Directive 2001/42/EC (see http://europa.

Box 12.2 Models of SEA Systems

EIA-based: SEA is implemented under EIA legislation, such as in the Netherlands, or carried out under separately administered procedures, such as in Canada and Hong Kong.

Environmental appraisal: SEA provision is made through a less formalized process of policy and plan appraisal, as in the UK prior to the European SEA Directive.

Dual-track system: SEA is differentiated from EIA and implemented as a separate process, such as the Netherlands 'Environmental test' or E-test of legislation and plans.

Integrated policy and planning: SEA is an integrated component of policy and plan development and decision-making, such as in New Zealand or for forest management plans in Saskatchewan.

Sustainability appraisal: SEA elements as separate evaluation and decision tools are replaced by integrated environmental, social, and economic assessment and appraisal of policy and planning issues, such as the former Australian Resource Assessment Commission and current sustainability plans in the UK.

Source: Based on UNEP (2002).

---

eu.int/comm/environment/eia/home.htm). In the Czech Republic, SEA was introduced by way of a reform to formal EIA legislation, whereas in Australia SEA was at first adopted informally as part of resource management programs and then in 1999 was introduced through formal legislated requirements (Noble, 2003).

In Canada, SEA is a policy requirement under the *1999 Directive on the Environmental Assessment of Policy, Plan and Program Proposals.* The *Directive* requires that environmental factors be considered for all departmental policy, plan, and program initiatives submitted to cabinet. Canada is recognized as a country that has made significant contributions to the development of environmental assessment above the project level (Table 12.1). The underlying principle of SEA in Canada is that in order to make informed decisions in support of sustainable development, decision-makers at all levels must integrate social, economic, and environmental considerations with policy, plan, and program development (see Chapter 2 and Noble [2004] for an overview of SEA development in Canada, and the Canadian Environmental Assessment Agency Web site <www.ceaa.gc.ca> for the cabinet directive on SEA).

Aims and Objectives

SEA is based on the premise that project-based EIA, which in essence reacts to a proposed project development, is not alone sufficient in ensuring the sustainable development of the environment. Consistent with the principles of Agenda 21, a series of actions aimed at greater sustainability and adopted at the 1992 Earth Summit by Canada and 177 other countries, SEA attempts to integrate *environment* into higher-order decision-making processes. Early integration of environment provides for a more proactive assessment process where alternatives are identified and assessed at an

Table 12.1   Brief Timeline of SEA Development in Canada

1990	Policy reform on environmental assessment; Canadian Environmental Assessment Research Council released guidelines for environmental assessment of policy and program proposals
1991	FEARO released *Environmental Assessment in Policy and Program Planning: A Sourcebook*
1992	North American Free Trade Agreement Environmental Review
1993	Natural Resources Canada introduced guidelines for the integration of environmental considerations into energy policies
1995	Canada released *SEA: A Guide to Policy and Program Officers*
1996	Environmental assessment of the new *Minerals and Metals Policy*
1997	Department of Foreign Affairs tabled *Agenda 2000* outlining its commitment to environmental reviews of recommendations submitted to cabinet
1999	Update to the 1990 cabinet directive released, and guidelines for implementation; SEA of Canada's commitment to Kyoto Protocol on greenhouse gas emissions

early stage in the policy and planning process before irreversible decisions are taken (Box 12.3). Thus, SEA is simply a way of ensuring that downstream project planning and development occurs within the context of the *desirable* outcomes that society wants to achieve. In this way, SEA addresses the *sources* rather than the *symptoms* of environmental change.

Box 12.3   SEA Benefits

Streamlining project EIA by early identification of potential impacts and cumulative effects.

Allowing more effective analysis of cumulative effects of broader spatial scales.

Facilitating more effective consideration of ancillary or secondary effects and activities.

Addressing the causes of impacts rather than simply treating the symptoms.

Offering a more proactive and systematic approach to decision-making.

Facilitating the examination of alternatives and the effects of alternatives early in the decision process before irreversible decisions are taken.

Facilitating consideration of long-range and delayed impacts.

Providing a suitable framework for assessing overall, sector-, and area-wide effects before decisions to carry out specific project developments are made.

Providing focus for project EIA on how proposed actions fit within the broader context of the region.

Continued

> Verifying that the purpose, goals, and direction of a proposed plan or initiative are environmentally sound and consistent with broader policy, plan, and program objectives for the region.
>
> Saving time and resources by setting the context for subsequent regional and project-based EIAS, making them more focused, effective, and efficient.
>
> Sources: Clark (1994); Cooper (2003); Kingsley (1997); Noble (2000); Sadler and Verheem (1996); Sadler (1998); Wood and Dejeddour (1992).

Types of SEA

The specific form of SEA will vary depending on the regulatory system and assessment context, but in general three types of SEA can be identified (Therivel, 1993).

(1) Often referred to as 'indirect SEA', **policy-based SEA** applies to policies that have no explicit 'on-the-ground' dimension, such as fiscal policies or national energy policies. The SEA of policies and legislative proposals is arguably the most significant type of SEA, as large-scale government policies commonly have more far-reaching effects than individual development plans, programs, and projects. A number of policy SEA applications do exist, but their number is limited in comparison to sector-based SEAs. Examples include the SEA of the North American Free Trade Agreement (see Hazell and Benevides, 2000), Canadian minerals and metals policy SEA (see Noble, 2003), and the SEA of two Danish bills under Denmark's EA system, one to amend laws relating to tenancy and housing conditions and rent subsidies, and the second a subsidy scheme for private urban renewal (see Elling, 1997).

(2) **Sector-based SEA** applies to sector-based initiatives, plans, and programs, such as forestry plans or oil and gas programs. The World Bank (1999) defines sector-based SEA as:

> an instrument that examines issues and impacts associated with a particular strategy, policy, plan or program for a specific sector; including the evaluation and comparison of impacts against those of alternative options and recommendation of measures to strengthen environmental management in the sector.

Emphasis is placed on the initiatives of and alternatives to particular sector-based plans or programs that may lead to environmental change (Box 12.4). Initiatives and their alternatives are evaluated within the context of sector-based objectives, existing environmental conditions, and current and proposed plans and priorities. Sector-based SEAs are typically defined by jurisdictional or sector-based boundaries, such as the extent of sector activities or the area subject to a particular sector-based plan (e.g., forest harvest plan).

A number of sector-based SEAs have been undertaken in recent years, particularly in the offshore oil and gas industry, for example: the Canada–Nova Scotia SEA of offshore licensing parcels and the SEA of the Laurentian sub-basin. While Canada, Norway, and Brazil all have systems for SEA offshore, perhaps the most advanced sys-

---

Box 12.4   Benefits of Sector-based SEA

- Avoids the limitations of project-specific EIAs in particular sectors in addressing broader environmental, social, and economic concerns.

- Prevents or avoids significant environmental effects through the assessment and development of sector-wide plans and programs before individual project decisions are taken.

- Provides opportunity for consideration of more effective or efficient sector plans or programs.

- Allows for planning and development of sector-wide environmental management strategies.

- Facilitates the inclusion of other sector interests in sector-wide planning and development.

- Provides a framework for considering the cumulative environmental effects of sector-wide development.

Source: World Bank (1999).

---

tem exists under the UK Department of Trade and Industry, where numerous SEAs for offshore licensing have been completed in recent years (Box 12.5).

Box 12.5   Department of Trade and Industry's Sector-based SEA, Offshore UK

Oil and gas exploration and production in the UK is regulated through a petroleum licensing system. A production licence grants exclusive rights to the holder to search for, drill for, and extract petroleum in specified areas. The UK Department of Trade and Industry (DTI), prior to the UK SEA Directive coming into force, adopted a policy decision that SEA will be undertaken prior to future wide-scale licensing of the continental shelf, and in 1999 began a series of sector-based SEAs to consider the implications of licensing for oil and gas activities. The first offshore SEA (SEA 1) was conducted in 1999–2000, with SEA 2 and SEA 3 following in 2001 and 2001–2. Currently, SEA 4, which commenced in late 2002, is in the stage of public consultation.

   The recently completed SEA 3 focused on parts of the central and southern North Sea, particularly 362 blocks, of which 30 are licensed, 205 have been previously licensed but now are relinquished, and 127 that have not previously been licensed. The proposed initiative under consideration for the region was to offer additional production licences for blocks in the UK sector through the next round of offshore licensing. The alternatives to the proposed initiative were not to offer any blocks for additional licensing, to license only a restricted area, or to stagger the timing of licensing activity in the area.

Continued

The SEA considered a number of objectives and issues, including environmental protection objectives and standards established for the area through existing policies and plans; existing environmental problems that may be exacerbated by the proposed initiative; and likely impacts of the initiative and its alternatives, including potential incremental, cumulative, and synergistic impacts.

An initial scoping process with academic and conservation organizations commenced in early 2001. This was followed by a broader consultation exercise to identify main components and areas of concern related to the proposed initiative and assessment. A range of issues was identified, from socio-economic and environmental concerns to coastal defence and opportunities for co-ordinating offshore wind farm activity. The initiative and proposed alternatives were assessed based on an initial environmental interaction matrix, expert judgement, stakeholder dialogue, and lessons from SEA 1 and SEA 2. Incremental effects were considered with regard to licensing actions likely to act additively with initiatives from other oil and gas activities, notably simultaneous and sequential surveying in existing and previously licensed areas. Cumulative effects were considered with regard to the potential impact of the initiative and alternatives on regional components in combination with the effects of other activities in the region of concern, notably seismic surveying and shipping on fishing activities and marine resources.

Based on the SEA output and recommended mitigation measures, it was recommended that DTI proceed with the licensing initiative as proposed.

Key Source/ Effect	Option 1: Not To Offer Any Blocks	Option 2/3: Restrict the Area Temporally or Spatially	Proposed Option: Proceed with Licensing as Proposed
Noise, seabed damage, physical presence, discharges, emissions, waste to shore, accidents, cumulative effects, transboundary effects	No benefit or disadvantage	Potential but minor environmental effect or socio-economic disadvantage. No clearly identifiable or justifiable seasonal or spatial restrictions identified.	Potential, but minor environmental effect or socio-economic disadvantage. No major effects are predicted given existing regulatory controls and mitigation.
Socio-economic effects	Potential but minor environmental effect or socio-economic disadvantage	Some benefit	Some benefit
Wider policy objectives	Potential significant environmental effect or socio-economic disadvantage	Strong benefit	Strong benefit

Source: Based on UK DTI (2002).

(3) Often referred to as **regional-based** SEA, spatial SEA includes regional plans and programs, such as land-use planning, which may include multiple sectors. The World Bank (1999) defines regional-based SEA as:

> an instrument that examines the environmental issues and impacts associated with a strategy, policy, plan for program for a particular region; including the evaluation and comparison of impacts against those of alternative options and recommendation of measures to strengthen environmental management in the region.

The purpose of regional SEA is to assess the impacts of plan and program initiatives within a particular region, in combination with other regional activities, in order to identify the preferred regional-based environmental planning or development strategy. The objective is to assist decision-making by systematically identifying a preferred option for regional management and development (Box 12.6). For example, the Bow River Valley Corridor regional study in Alberta, while not explicitly labelled as an SEA, placed considerable attention on evaluating the potential impacts of alternatives and competing land uses, including oil and gas projects, mining development, urban expansion, and agricultural grazing in the Bow River Valley. The outcome of the assessment assisted in identifying preferred land-use development patterns and in setting limits to further development.

Regional SEA is often defined by environmental or ecological boundaries, may include multiple sectors, and is typically driven by environmental planning or management initiatives, state-of-environment reports, or the initiatives of and stresses caused by multiple sectors—an example of which is the Northwest Territories Cumulative Effects Assessment and Management (CEAM) Framework (Box 12.7).

Box 12.6 Benefits of Regional-based SEA

- Influences planning in the area.

- Encourages a long-term perspective on regional planning.

- Collects and organizes regional environmental data.

- Allows for comprehensive planning of region-wide environmental monitoring and management.

- Provides a basis for planning and decision-making across sectors and administrative boundaries.

- Helps identify and avoid conflicting developments and plans.

- Sets the context for and strengthens subsequent sectoral and project-based assessments.

- Provides a framework for the CEA of multiple plans and activities in a particular region.

Source: World Bank (1999).

Box 12.7 CEA Regional Strategic Framework for the NWT

Following development of the Diavik Mine project in the Northwest Territories, Canada's second diamond mine, the federal Ministers of Environment and Indian Affairs and Northern Development initiated a plan to develop a NWT Cumulative Effects Assessment and Management (CEAM) strategy and framework, as well as a generic regional strategic management and assessment framework for the North. The goal of the CEAM strategy and framework is to provide a systematic and co-ordinated approach to the assessment and management of cumulative effects in the NWT. In doing so, several key components are identified by the CEAM strategy and framework as necessary to manage cumulative environmental effects, including a thorough CEA of all proposed projects, a regional steering body representing affected and interested groups, a regional ecosystem health monitoring program, land-use and watershed planning, accessibility of information, and a guiding research body.

The overall purpose of the CEAM strategy and framework is to facilitate:

- mechanisms to make informed decisions about the environment and human activities;
- systematic and co-ordinated approaches to CEA and management;
- common principles and links between environmental management processes, land-use planning, land and water management, cumulative impact monitoring, and environmental auditing;
- improved certainty, efficiency, transparency, and accountability of processes.

Some of the key objectives of the CEAM framework from a regional assessment perspective include:

- providing greater certainty regarding how cumulative effects of development will be considered in planning and assessment;
- facilitating more effective and efficient review of project assessments, including the reduction of real or perceived duplication in cumulative effects assessment;
- developing regional approaches to adaptive management;
- identifying and addressing any gaps, omissions, or research needs;
- developing approaches to integrating traditional knowledge into cumulative effects assessment and management;
- building partnerships and integrating communities into cumulative effects assessment and management;
- contributing to a better understanding of the environmental and socio-economic systems in the NWT, including a better predictive capacity for assessing changes;
- developing and agreeing on regional environmental thresholds, carrying capacities, and acceptable levels of landscape change to support decision-making and project development.

Source: Based on <www.ceamf.ca/>.

Another example of a regional SEA, which is currently underway, is the regional environmental study of the Great Sand Hills, Saskatchewan. The Great Sand Hills are located in the southwest part of Saskatchewan and cover approximately 2,000 square kilometres of prairie. The Great Sand Hills contain one of the most significant concentrations of active and inactive sand dunes of any populated area in Canada. Rich in ecological diversity and rare and endangered species, the Great Sand Hills are a growing tourist attraction and lie on top of significant reserves of natural gas. In early 2005 the government of Saskatchewan appointed an international advisory committee to undertake an environmental study of land use in the area. The purpose of the study is to identify the cumulative environmental effects of current and potential future land use and to identify sustainable land-use options for future planning, development, and ecological protection. The study is expected to be completed by 2007. Information on the study as it unfolds can be found at <www.se.gov.sk.ca/GSH/>.

SEA FRAMEWORKS

As suggested earlier, SEA extends EIA upstream but at the same time it adopts a different set of principles. Strategic environmental assessment has advanced considerably in recent years, including the development of new SEA frameworks. There is no single agreed-upon framework for SEA, but based on recent international experiences, a number of generic steps can be identified for SEA implementation. It is important to note that, similar to EIA, specific design requirements are often necessary within each application—including specific methods and techniques. Most of the methods and techniques required for SEA are readily available from project-level assessment and policy appraisal. A brief outline of a generic SEA framework, including selected examples of methods and techniques, is presented in the following sections. Much of the framework is modelled after EIA terminology and thus not explained in any great detail. The reader is encouraged to refer to the previous chapters and to the Glossary for a refresher of terms and concepts.

Screening
Screening involves determining whether an SEA is needed. Requirements and triggers for SEA vary considerably from one nation to the next, but there are essentially three types of provisions for SEA in international practice:

- legislative or mandatory provisions, such as in Western Australia;
- administrative orders, such as in Denmark and Canada;
- advisory or operational policy provisions, such as provided for by the World Bank, which suggests a more discretionary approach to SEA.

Scoping
At least four questions should be asked in relation to potential effects of a proposed policy, plan, or program:

- What are the objectives of the SEA?
- What are the objectives of the strategic initiative or proposed plan?
- What is the baseline condition?
- What are the strategic alternatives?

SEA objectives. Scoping the SEA involves determining the function or role of the SEA by asking:

- What is the purpose of the assessment?
- What is hoped to be accomplished by the assessment?

In sector-based SEA, such as for offshore oil and gas, the assessment objectives may be more closely related to industry objectives for environmental performance, business initiatives, or focus on industry and industry-related activities within the region (Box 12.8). For a broader regional-based SEA, such as the development of a regional land-use plan, the assessment objectives may focus on identifying an appropriate direction for longer-term sustainable regional development, including the implications of multiple-sector developments, including housing, oil and gas exploration, and transportation, and interactions between these activities within the region.

Establish plan objectives. Scoping the plan objectives involves developing a reference framework and defining the nature and scope of the initiative (the particular need or demand) or proposed plan or program, specifically: What are the objectives of the proposed plan or planning initiative?

Identify alternatives. Unless there is more than one potential and feasible way to proceed, there is no decision choice to be made and therefore no SEA is required; this includes the option of no plan or action. Alternatives should include:

- the baseline condition of no initiative or no plan or program implementation;
- the proposed initiative, plan, or plan or program implementation procedure;
- alternatives to the initiative, proposed plan, or plan or program implementation procedure.

This might include alternatives that emphasize different:

- spatial and temporal attributes or option implementation;
- modes or processes, including technologies or methods;
- means of meeting the plan or strategic initiatives;
- different objectives.

Alternatives should only be considered that are:

- consistent with the initiative or planning goals and objectives;
- consistent with broader regional or sectoral environmental goals and objectives;
- not in conflict with existing regional or sector policies, plans, or activities; and/or
- economically, technologically, and institutionally feasible.

The objective is to identify alternatives that are 'more sustainable', 'least negative', 'more in line' with existing policies, plans, programs, and projects, or that trigger the 'least significant' amount of environmental change. These alternatives should be selected in light of the specified plan or program and assessment objectives identified

---

Box 12.8 Mackenzie Delta and Beaufort Sea Hydrocarbon Reserves SEA

Due in part to an upswing in market conditions, since 1999 there has been a growing interest in exploration and development of oil and gas reserves in the Mackenzie Delta and Beaufort Sea regions of northern Canada. A large portion of the Delta is also home to the Kendall Island Bird Sanctuary, which itself sits on top of economically valuable oil and gas reserves. Due to concern over the likelihood and nature of potential development, and the presence of the bird sanctuary, Environment Canada and its Canadian Wildlife Service conducted an SEA for the region. The objective of the SEA was to identify an appropriate strategic initiative, in this case a plan for oil and gas development. More specifically, the objective was to consider and assess options for managing oil and gas activity in the Kendall Island Bird Sanctuary. Included within the broader SEA objective of 'plan development' were two principal sub-objectives:

- to identify opportunities to incorporate sustainable development approaches into the development of a management policy for the region;
- to identify opportunities to explore and potentially extract petroleum reserves while minimizing the environmental footprint.

Four options were assessed to identify an appropriate strategy for managing oil and gas development in the region:

1. discontinue any development in the sanctuary;
2. allow development to continue as it has in the past, assessing projects on a case-by-case basis;
3. delist or reconfigure the sanctuary to allow developments to occur outside sanctuary boundaries;
4. manage oil and gas in the sanctuary in a manner that would minimize the developmental footprint by pursuing a memorandum of agreement with industry.

The fourth option was identified as the preferred strategic direction for the region.

---

during the screening and scoping phases. If, in the case of a proposed plan or program, a strategic alternative is likely to cause a 'more significant negative effect' than the proposed plan or program itself, then it should not be considered.

Establish baseline condition. Establishing the baseline conditions involves asking:

- What are the VECs of concern?
- What is the current (and cumulative) condition of potentially affected VECs?
- What effects or issues are likely to result from each alternative?
- What are the thresholds of concern or condition objectives associated with the VECs?
- What are the appropriate spatial and temporal boundaries for the type of region or sector under consideration?

Only those environmental or socio-economic components likely to be receptors of effects that are important to ecological functioning or valued by society should be considered, including regional air quality, water resources, habitats, spaces of cultural significance, social well-being, or, at the strategic level, an existing policy or plan.

Scope the alternatives. Scoping the alternatives should be done after the identification of VEC receptors, but before the establishment of VEC criteria or objectives. The focus should be on identifying broad, conceptual issues rather than on detailed impacts and interactions per se. The objective at this stage is to identify those alternatives suitable for further consideration. For each alternative, the following minimum factors should be considered in order:

- Is the alternative compatible with the goals and objectives of existing and proposed policies, plans, and activities for the sector or region?
- Is there a potential interaction (either positive or negative) between activities associated with the proposed alternative and existing VECs?
- Are potentially affected VECs already affected (either positive or negative) by existing policies, plans, or activities in the sector or region?
- Will the alternative have an effect on current VECs (either positive or negative) in combination with already existing and proposed policies, plans, and activities?

An example of how options might be scoped at an early stage in light of potential interactions with existing regional or sector plans and activities is depicted in Table 12.2.

Establish VEC criteria. Similar to predicting project-based impacts, making useful conclusions about effects requires some 'limit of change' or 'specified objective' to which the incremental effects of the plan can be assessed. Thus, establishing criteria for the identified VECs is critical to evaluating the environmental effects of the initiatives, proposed plan, or plan or program options and alternatives. These criteria represent the specific parameters, guidelines, or standards that must be met, such as a carrying capacity or a set limit of environmental change, and are usually set in the context of a broader environmental vision for the region or sector, and based on VEC indicators. As discussed Chapter 5, there are three principal approaches to establishing VEC criteria or objectives:

1. absolute ecological or socio-economic thresholds;
2. acceptable limits of change;
3. desired conditions or outcomes.

Delineate assessment boundaries. Spatial and temporal boundaries provide a frame of reference for SEA and determine the level of analysis. Spatial and temporal bounding for SEA is no different, in principle, than that for project-based EIA. The primary difference at the strategic level concerns the nature of the activities that must be considered. For example, geographic relationships, common resources, and proposed activities must be viewed not only from the perspective of physical and socio-economic phenomena, but also from the perspective of current and proposed policies, plans, or programs that may interact with the proposed plan or program.

Table 12.2   Simple Checklist of a Preliminary Alternatives Scoping Matrix

Plan options or alternatives	Is the alternative compatible with existing and proposed policies, plans, and activities?	Is there a potential for interaction with and effect on current VECs?	Is the VEC already affected by existing plans or activities?	Is there a potential for cumulative interaction?	Should the alternative be given further consideration?
A1		water quantity			☐ Yes
		hunted species			☐ No
		potable water			☐ Uncertain
A2		water quantity			☐ Yes
		hunted species			☐ No
		potable water			☐ Uncertain
A3		water quantity			☐ Yes
		hunted species			☐ No
		potable water			☐ Uncertain

Source: Based on the UK Office of the Deputy Prime Minister (2003).

Assessing Initiatives and Option Impacts

Only those options or alternatives that are compatible with the goals and objectives of existing and proposed policies, plans, and initiatives for the sector or region should be considered for detailed assessment. This way, the focus of assessment is on the *desirability* of the initiatives with regard to VEC objectives, rather than on trying to resolve conflicting goals and objectives between proposed and existing initiatives.

The assessment of options should include consideration of whether:

- the proposed option will have an effect on specified VECs;
- there will be potential cumulative effects on valued resources;
- other policies, plans, or actions may affect the same resources;
- whether the effects are likely to be significant.

These should be considered within the context of interactions with and the contributions of the cumulative effects of current and reasonably foreseeable policies, plans, and initiatives within the sector or region with or without the proposed plan or initiative.

Identify cause-effect relationships for VECs. At the strategic level, particularly when dealing with regional-based assessments, specific cause-effect relationships between proposed options and VECs may be difficult to establish. The objective is not to spend a great deal of time and resources in identifying ecosystem-wide linkages, for example, but rather to gain a general sense of understanding of important components and possible interactions.

Determine environmental changes likely to affect VECs. The analysis of potential effects involves prediction of the potential impacts of each option on the VECs by identifying the primary sources of stress and associated cumulative effects pathways. This is a formidable task, and the levels of complexity and uncertainty increase as one moves from projects to plans and from sectors to regions. The objective is to identify, in a general sense, potential stressors and VEC responses. The level of detail depends on the availability of data. One approach is to conceptualize the cause-effect-response relationship using *network analysis* (Figure 12.1).

Identify VEC receptor response. This requires understanding of the present conditions of VECs and predicting how each might react based on the specified VEC indicators. In short, the objective is to capture, based on the causal networks established, how the VEC receptor might respond or deviate from its current condition given the multiple interactions and pathways with and without each of the proposed options. Depending on the nature of the VEC receptor and level of information available, it may be possible to model and identify trends over time based on different scenarios of cause-effect relationships for each option; in other cases, expert-based forecasting, such as the Delphi approach, might be best suited. In general, such methods can be readily adopted from EIA-driven approaches (Box 12.9).

Assess alternatives against VEC objectives. For SEA, the issue may not always be about 'predicting' and 'mitigating' impacts so much as it is about identifying preferred directions and outcomes. In other words, what is required to achieve a desired future or VEC outcome, and what are the consequences of different choices? VEC or

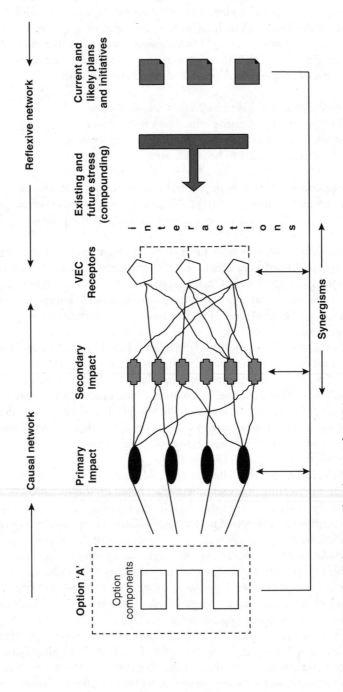

Figure 12.1 VEC-based Network Diagram for SEA

Box 12.9 Selected CEA Methods and Techniques Suitable to SEA of Plan Options

Impact models: flow diagrams; loop diagrams; network diagrams; quantitative modelling

Geographic information systems: spatial relationships; scenarios over time and space

Trend analysis: temporal change

Expert judgement and valuation: forecasting; Delphi approaches; weighting and scoring

Econometric techniques: monetary evaluation; benefit-cost analysis

Checklists and matrices: lists of option impacts; categorization of impacts

Risk assessment: probability of outcomes; scenario analysis

---

plan goals and objectives, such as minimizing the potential for effects and cumulative interactions between plans or programs affecting water quality, for example, and alternative means of achieving those goals and objectives are evaluated. Thus, emphasis might be placed on assessing the feasibility and desirability of meeting a desired target or on selecting alternatives that minimize potential cumulative effects or maximize environmental and socio-economic contributions.

For each VEC objective, one should consider how *desirable* is strategic option '*i*' compared to option '*j*' when taken into consideration with the effects of current and reasonably foreseeable policies, plans, and initiatives. How the options are evaluated in light of the objectives again depends on the availability of data and desired decision resolution and may include, for example, expert-based judgement or more complex *modelling* and *simulation analysis*. In keeping with the example and selected techniques discussed in Chapter 11, paired comparison weightings, an example of alternatives assessment based on expert judgement is depicted in Box 12.10.

Box 12.10 Example of Alternatives Assessment Based on Simple Paired Comparison Weighting

(Refer to Chapter 11, Box 11.3, for an explanation of the paired comparison technique.)

How desirable is strategic option '*i*' compared to option '*j*' when taken into consideration with the effects of current and reasonably foreseeable policies, plans, and initiatives on the specified VEC?

VEC objective '*n*'	Option 1	Option 2	Option 3
Option 1	1	1/3	7
Option 2	3	1	7
Option 3	1/7	1/7	1

Continued

Where:

9 = option 'i' is extremely more desirable in comparison to option 'j'

7 = option 'i' is strongly more desirable in comparison to option 'j'

5 = option 'i' is moderately more desirable in comparison to option 'j'

3 = option 'i' is slightly more desirable in comparison to option 'j'

1 = option 'i' is equally desirable to option 'j'

Compared to option 'j', option 'i' is more likely to: a) meet the VEC objective and/or b) minimize overall effects on the VEC.

1/3 = option 'i' is slightly less desirable in comparison to option 'j'

1/5 = option 'i' is moderately less desirable in comparison to option 'j'

1/7 = option 'i' is strongly less desirable in comparison to option 'j'

1/9 = option 'i' is extremely less desirable in comparison to option 'j'

Compared to option 'j', option 'i' is less likely to: a) meet the VEC objective and/or b) minimize overall effects on the VEC.

Normalized matrix:

VEC objective 'n'	Option 1	Option 2	Option 3
Option 1	0.24	0.23	0.47
Option 2	0.72	0.68	0.47
Option 3	0.03	0.10	0.07

Priority vector : Option 1 = 0.31; Option 2 = 0.62; Option 3 = 0.06

$$0.31 \begin{bmatrix} 1 \\ 3 \\ 1/7 \end{bmatrix} + 0.62 \begin{bmatrix} 1/3 \\ 1 \\ 1/7 \end{bmatrix} + 0.06 \begin{bmatrix} 7 \\ 7 \\ 1 \end{bmatrix} = \begin{bmatrix} 0.94 \\ 1.97 \\ 0.20 \end{bmatrix}$$

Ranking of alternatives based on VEC objective 'n' (divide the vector by original priorities)

Option 1 = 0.94 / 0.31 = (3.03)

Option 2 = 1.97 / 0.62 = (3.18)

Option 3 = 0.20 / 0.06 = (3.33)

Option 3 (3.33) > Option 2 (3.18) > Option 1 (3.03)

Normalize the ranking by $[(i - i_{min}) / (i_{max} - i_{min})]$ to display the relative 'magnitude' of the ranking for each VEC objective:

(Option 1 = 0; Option 2 = 0.50; Option 3 = 1)

Option 1		Option 2		Option 3
0	0.25	0.5	0.75	1.0

Determine 'cumulative desirability' of the option:

Repeat the above procedures for each VEC objective. The 'cumulative desirability' and final preferred option or initiative is determined by summing the normalized rankings across all VEC objectives. For example:

Continued

Objectives	Option 1	Option 2	Option 3
VEC objective 1	0.00	0.50	1.00
VEC objective 2	0.00	1.00	0.24
VEC objective 3	0.36	0.00	1.00
VEC objective 4	0.11	0.00	1.00
'Cumulative desirability'	0.47	1.50	3.24

Normalized: *Option 1* = 0; *Option 2* = 0.37; *Option 3* = 1

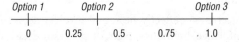

The net result is a ranking of all options and identification of the preferred option based on the (unweighted) VEC objectives. To determine the weighted rankings, simply multiply the option scores by VEC weights (importance) prior to normalizing.

Determining Significance

A number of questions can be examined across VECs to determine whether effects associated with strategic initiatives are significant, notably:

- Will the proposed initiative or plan option generate effects that will exceed thresholds as defined in regulations, guidance, or as established by maximum allowable effects levels (see Chapter 9)?
- Are the effects significant when considered in light of the current conditions of the affected VEC?
- Will the plan or plan option generate effects on VECs that are likely to interact cumulatively with other plans or activities in the region on already stressed VECs?
- Will the effects of the plan or plan option be permanent?
- Can the potential effects be mitigated?
- Does the initiative or plan option conflict with the goals and objectives of existing plans and activities?

Identifying the Preferred Option

The outcome of the assessment indicates a desirable strategic direction. In other words, when compared to all other proposed options, a preferred strategy for action is identified based on its:

- potential effects on specified VECs;
- potential for cumulative effects on valued resources;
- interactions with and the contributions of current and reasonably foreseeable policies, plans, and initiatives within the sector or region, with or without the proposed plan or initiative;
- specific VEC objectives.

Follow-up
No assessment process is complete without a follow-up or monitoring component. There are three broad types of follow-up and evaluation in SEA:

1. **Input evaluation** concerns SEA procedures and requirements, such as SEA purpose and objectives, data quality, and linkages to policy, planning, or program initiatives.

2. **Process evaluation** concerns SEA application, such as validity of methods, nature of alternatives, impact identification and analysis, recommended management measures, follow-up programs, transparency, participation, efficiency, and usefulness of SEA findings.

3. **Output evaluation** concerns SEA effectiveness, and involves determination of whether the analysis informed the policy, plan, or program decision process, whether the proposed policy, plan, or program was modified as a result of the SEA, and whether what were intended as a plan or initiative outcomes were in fact realized. In this sense, follow-up may focus on identifying:

   - changes in environmental conditions that are within current acceptable limits and reversible (suboptimal);
   - no change in pre-project environmental conditions (status quo); or
   - improvements in pre-project environmental conditions (optimal).

DIRECTIONS IN SEA

Strategic environmental assessment is still quite new and relatively limited in terms of its adoption. Noble (2004), for example, in a survey of SEA administrators in Canada, identified a number of barriers to SEA development and implementation including, in descending order or importance:

- problems with co-ordination/responsibilities;
- lack of methodological/procedural guidelines;
- lack of training and expertise;
- lack of methods and techniques;
- financial constraints;
- insufficient legislative requirements;
- lack of agreement concerning benefits of/need for SEA.

Nonetheless, SEA is advancing and new and innovative approaches to policy, plan, and program assessment are being developed. SEA can do much more than simply expand EIA upstream; it is an important tool towards ensuring sustainability by considering environment as an integral component of policy, plan, and program decision-making.

KEY TERMS

input evaluation
output evaluation
plan
policy
policy-based SEA

process evaluation
program
regional-based SEA
sector-based SEA
strategic environmental assessment

STUDY QUESTIONS AND EXERCISES

1. What are the potential benefits and challenges to applying environmental assessment to policy?
2. Identify the different types of SEA, and provide an example of the type of problem or situation to which each might apply.
3. What are the provisions, if any, for policy, plan, or program assessment in the political jurisdiction in which you live? What do you see as the main challenges to SEA where you live?
4. Visit the Web sites for the Canadian cabinet directive for implementing SEA at <www.ceaa.gc.ca> and the European Directive at <http://europa.eu.int/comm/environment/eia/home.htm>. Compare and contrast the directives in terms of their objectives, scope, requirements for reporting, and public involvement.
5. An important part of SEA is the identification of alternatives. One question that emerges from this, however, is who should be involved. Consider two hypothetical plan proposals in your area, one for oil and gas licensing and one for the development of a regional land-use plan. Discuss who should be involved in the identification of alternatives.
6. Obtain a completed SEA from your local library or government registry, or access one on-line. Using Box 12.1 as a guide, explore the SEA for evidence of 'strategic' characteristics. In other words, are the characteristics listed in Box 12.1 present in the SEA? Are there certain characteristics that seem to be missing? Compare your findings to those of others.

REFERENCES

Clark, R. 1994. 'Cumulative Effects Assessment: A Tool for Sustainable Development', *Environmental Impact Assessment Review* 12, 3: 319–22.

Cooper, L. 2003. *Draft Guidance on Cumulative Effects Assessment of Plans.* EPMG Occasional Paper 03/LMC/CEA. London: Imperial College.

Elling, B. 1997. 'Strategic Environmental Assessment of National Policies: The Danish Experience of a Full Concept Assessment', *Project Appraisal* 12, 3: 161–72.

Hazell, S., and H. Benevides. 2000. 'Toward a Legal Framework for SEA in Canada', in M.R. Partidario and R. Clark, eds., *Perspectives in Strategic Environmental Assessment.* New York: Lewis Publishers.

Kingsley, L. 1997. *A Guide to Environmental Assessments: Assessing Cumulative Effects.* Hull, Que.: Parks Canada, Heritage Canada.

Noble, B.F. 2000. 'Strategic Environmental Assessment: What Is It and What Makes It Strategic?', *Journal of Environmental Assessment Policy and Management* 2, 2: 203–24.

———. 2003. 'Auditing Strategic Environmental Assessment in Canada', *Journal of Environmental Assessment Policy and Management* 5, 2: 127–47.

————. 2004. 'A State-of-Practice Survey of Policy, Plan, and Program Assessment in Canadian Provinces', *Environmental Impact Assessment Review* 24: 351–61.

Sadler, B. 1998. 'Ex-post Evaluation of the Effectiveness of Environmental Assessment', in Alan L. Porter and John J. Fittipaldi, eds., *Environmental Methods Review: Retooling Impact Assessment for the New Century*. Fargo, ND: The Press Club.

———— and R. Verheem. 1996. *Strategic Environmental Assessment: Status, Challenges and Future Directions*, Report 53. Ottawa: Canadian Environmental Assessment Agency.

Sheate, W., S. Dagg, J. Richardson, P. Wolmarans, R. Aschemann, J. Palerm, and U. Steen. 2001. *SEA and Integration of Environment into Strategic Decision-Making*, vol. 1. Report prepared for the European Commission under contract B4-3040/99/136634/MAR/B4.

Therivel, R. 1993. 'Systems of Strategic Environmental Assessment', *Environmental Impact Assessment Review* 13: 145–68.

UK DTI. 2002. *Strategic Environmental Assessment of Parts of the Central and Southern North Sea: SEA 3*. UK Department of Trade and Industry.

UK Office of the Deputy Prime Minister. 2003. *Strategic Environmental Assessment Directive: Guidance for Planning Authorities*. Practical guidance on applying European Directive 2001/42/EC.

United Nations Environment Program (UNEP), Economics and Trade Program. 2002. *Environmental Impact Assessment Training Manual*, 2nd edn. New York: UNEP.

Wood, C., and M. Dejeddour. 1992. 'Strategic Environmental Assessment: EA of Policies, Plans and Programmes', *Impact Assessment Bulletin* 10, 1: 3–23.

World Bank. 1999. *Environmental Assessment*. Operational Policy and Bank Procedures, no. 4.01.

World Commission on Environment and Development. 1987. *Our Common Future*. Oxford: Oxford University Press.

Alternative Means: An Exercise in Project Assessment

INSTRUCTIONS

The purpose of this exercise is to provide course participants with an opportunity to apply various EIA methods, techniques, and procedures discussed in this guide to a development problem that mimics a real-world situation. Students should work in groups (consulting teams) to prepare a preliminary scoping and assessment report, including a project assessment matrix, criteria and a protocol for determining impact significance, suitable techniques for predicting impacts, and management and impact monitoring mechanisms. The exercise is designed to be applied to the 'local environment' within which the course is being delivered, thus it is necessary that the students be provided with a copies of topographic and political maps of the region.

HOUSEHOLD AND LIGHT INDUSTRIAL WASTE SANITARY LANDFILL PROJECT

Background

Waste Management Inc. recently filed an application for the development and operation of a regional sanitary landfill site for household and light industrial waste. You have been contracted by the company to prepare a preliminary scoping and assessment report in preparation for a full environmental assessment of the proposed facility.

The proposed landfill project will consist of three main components:

- construction of a site access road;
- excavation and operation of an open-pit waste disposal site;
- project and site decommissioning.

Household and light industrial waste from regional communities will be shipped to the disposal site via highway transport. The construction of a single-lane, gravel-surfaced site access road capable of withstanding the load of the heavy equipment used to ship the waste material is required. Site development will require the excavation of a large open pit, to be lined with an impermeable polyurethane sheet to avoid hydraulic contact with groundwater. The landfill pit design is depicted in Figure A.1 below.

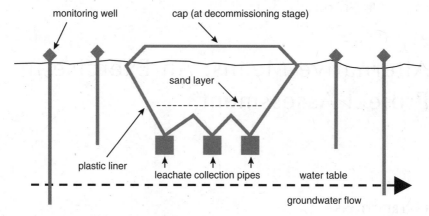

Figure A.1 Landfill Project Engineering Design

Total employment during construction is estimated to peak at 125 persons during the first three months of the project. It is anticipated that, on average, 70 per cent of these jobs will be filled by workers from the region, primarily for labour-intensive work associated with pit excavation and road construction. Analysis of the local labour force suggests that about 30 per cent of the peak labour force requirements will be provided by local residents. Construction is expected to last for eight months.

The design life of the facility is approximately 12 years, during which time the permanent on-site workforce will be about 25 persons, including site managers, site maintenance, and security. At the end of this period the disposal site will be decommissioned. This will involve covering the waste with compacted clay, a low-permeability layer of synthetic material, a drainage layer of stone and gravel, and a vegetated top. Monitoring wells will remain intact and operable.

The total cost of the project is estimated to be $26.5 million, including the access road and landfill construction and site decommissioning.

Report Requirements
As the hired consultant, you are asked to deliver a report detailing the following:

Location Analysis
Construct a general 'screening checklist' of location criteria for a sanitary landfill project (e.g., 50 metres from a waterway, minimum of five kilometres from residential area, etc.) and identify a potential project location on your map sheet.

Project and Baseline Description
A description of the project actions and current environmental baseline is required. This will consist of three components:

1. A list of project actions associated with the construction and operation of the sanitary landfill, including, for example, site clearing, pit construction, and waste transport.

2. Description of the human and biophysical baseline environment of the locality, including population, labour force statistics, existing environmental quality and stress.
3. Identification of the likely affected and interested publics for consultation.

VEC Identification
Identify potentially affected human and biophysical VECs, and delimit an appropriate spatial boundary for the assessment.

Impact Identification
The proponent requests that two impact identification components be included in the preliminary report:

• A simple network visually identifying the linkages and order of impacts associated with project actions and affected VECs.
• An impact matrix identifying and characterizing the potential impacts on the affected VECs. The nature of the impact matrix is at the discretion of the consultant, but must include (1) project actions and affected environmental components; (2) some indication (e.g., scale or weighting) of the potential impact significance, which should be clearly explained; (3) a one-sentence statement of each of the identified impacts.

Techniques for Impact Prediction
Once all impacts are identified, the proponent has requested that the consultant identify four of the most significant environmental impacts, of which at least one must concern the biophysical environment and one the human environment, and one must be a positive impact. For each impact, provide a maximum two-page description of the predictive technique(s) the proponent might use for impact prediction. A brief assessment of the strengths and limitations of the predictive technique should also be included.

Impact Management Measures
For each of the above impacts, recommendations should be made as to appropriate and feasible impact management and mitigative measures. Impact management measures for each impact should be detailed in one page.

Monitoring Components
A monitoring program is required in order to measure actual environmental change and to evaluate the effectiveness of impact management measures. For each of the four significant impacts:

• identify appropriate impact indicators for monitoring;
• identify the type of data that should be collected for each of these indicators and general location(s) where the data are to be collected.

Additional information and references, including a project location map, should be appended to the consulting report.

APPENDIX 11

Alternatives To: An Exercise in Strategic Assessment

INSTRUCTIONS

The purpose of this exercise is to provide course participants with an opportunity to apply various SEA methods, techniques, and procedures discussed in this guide to a policy and planning problem that mimics a real-world situation. Students should work in groups (consulting teams) to prepare an SEA report including an assessment matrix, criteria and a protocol for determining significance, and identification of the preferred strategic direction. The exercise is designed to be applied to the 'regional or national environment' within which the course is being delivered.

ELECTRICITY SUPPLY SOURCE ALTERNATIVES

Background
The demand for electricity within the region is projected to increase at an annual average rate of 1.8 per cent over the next 20 years. The current electricity supply system is expected to reach its capacity within the next two years, and a strategy is required now in order to plan for the future and appropriately address electricity demand projections.

A recent workshop involving government and various industry representatives identified a number of technologically feasible alternatives for addressing the projected increase in electricity demand, including:

Option 1: An intensive energy conservation and demand reduction program.
Option 2: The development of one new coal-fired generation station.
Option 3: An increase in renewable energy, namely wind power.
Option 4: An increase in hydroelectric generation capacity.
Option 5: The development of one new nuclear reactor.

Report Requirements
You have been contracted by the proponent, the state government, to provide an assessment of supply alternatives based on your expert judgement and consultation with other experts. You are asked to deliver a report detailing the following:

Assessment Factors

A list of assessment factors and criteria are required in order to assess the implications of each proposed alternative. Included in this list should be economic factors and objectives (e.g., minimize electricity cost to consumers), social factors (e.g., ensure equitable access and distribution), and environmental factors (e.g., minimize atmospheric emissions). Criteria should be limited to those seven to nine criteria that are deemed most important.

Criteria Weighting

The assessment factors are to be weighted in order to ascertain their relative importance in the decision and assessment process. Include in the report a description of the weighting procedure and a table of final weights for each criterion. Given that this is a preliminary assessment, it is requested by the proponent that weights be derived based on your expert judgement and experience.

Alternatives Assessment

Each alternative is to be assessed against the proposed criteria. Include in the report a description of the methods used for assessment, namely those that rely on your expert judgement and experience, and the necessary tables or impact matrices indicating the impacts or preferences for each alternative on the basis of the individual criterion.

Identification of Strategic Option(s)

Identify, based on the above assessment, an overall ranking of electricity supply options.

Sensitivity of Decision

Examine the sensitivity of the rankings and assessment outcomes by examining two critical scenarios of particular importance to the preferred electricity supply option. First, examine the implications of the development of a stringent emissions policy by increasing the weights assigned to any environmental factors and criteria by 50 per cent. Second, examine the implications of an increase in the cost of electrical generation by assigning a 50 per cent increase to those economic factors and criteria. Discuss the outcomes or implications of these changes to the ranking of supply options.

Additional information and references, including a project location map, should be appended to the consulting report.

Glossary of Key Terms

accuracy

The amount of bias applied to the system-wide impact predictions.

acid mine drainage

Drainage from surface mining, deep mining, or coal refuse piles, usually highly acidic with large concentrations of dissolved metals.

active publics

Publics that affect assessment decisions.

activity information

The effects possibly generated due to a project.

adaptive environmental assessment and management (AEAM)

A simulative modelling approach that quantifies and defines resources and environmental problems.

additive impacts

Individual minor actions that may be separate or related and that have a significant overall impact when combined.

ad hoc approaches

Applying EIA findings of past common projects to predict or understand proposed or existing projects.

adverse effects

Direct project effects that have potential to cause long-term irreversible undesirable environmental damage or change.

alternative means

Under the Canadian Environmental Assessment Act, 'alternative means' refers to different ways of carrying out the proposed project—typically alternative locations, timing of activities, or engineering designs.

alternatives to

Under the Canadian Environmental Assessment Act, 'alternatives to' refers to different ways of addressing the problem at hand or meeting the proposed project objectives; renewable energy, for example, would be considered an 'alternative to' a proposed coal-fired generating station.

ambient environmental quality monitoring

A pre-project assessment of the surrounding environment inclusive of biophysical and socio-economic factors; collected information is used as a baseline in comparing a project's environment during development, operation, and post-operation against unaffected control sites in order to monitor the impact of a project.

ambitious bounding

Selecting a project boundary that is too large, often leading to the proposed project seeming insignificant.

amplifying effects

Incremental effect additions where each increment has a larger effect than the one preceding.

analysis scale

The scale applied to examine VECs and impacts across space.

antagonistic impacts

Individual adverse impacts that have potential to partially cancel out when combined together.

assimilative capacity

The ability for an environment to incorporate pollutants into its system without adverse effects to is natural state.

auditing

An objective examination or comparison of observations with predetermined criteria.

balance model

Model designed to identify inputs and outputs for specified environmental components; these models are commonly used to predict change in environmental phenomena.

baseline condition

A description of pre-project environment, which is inclusive of the cumulative effects of previous activities and the future environment in absence of the proposed project.

baseline study

Description of the biophysical and socioeconomic state of the environment at a given time that can later be used for the comparison of environmental change through time.

biophysical environment

Consists of air, water, land, plants, and animals.

buffer zone

An area of undisturbed environment; a commonly used mitigation practice.

Canadian Environmental Assessment Act

The legal basis for federal EIA in Canada; sets out responsibilities and procedures for EIA of projects that involve federal authorities; introduced to Parliament in 1992, proclaimed in 1995, and most recently revised in 2003.

Canadian Environmental Assessment Agency

The federal agency created in 1994, replacing the Federal Environmental Assessment Review Office, to oversee Canadian federal EIA and implementation of the Canadian Environmental Assessment Act.

case-by-case screening

Evaluating project characteristics against a checklist of regulations and guidelines.

checklists

Method used to create a comprehensive list of effects or indicators of environmental impacts that a project might generate.

class screening EIA

An assessment method used to streamline projects deemed unlikely to cause adverse environmental effect or that are considered routine undertaking.

closed scoping

EIAs with content and scope predetermined by law, where modifications can only be made through closed consultation between the proponent and the responsible authority or regulatory agency.

compensation

The measures taken by the proponent to make up for adverse environmental impacts of a project that exist after mitigation measures have been implemented.

compliance monitoring

Monitoring to ensure that all project regulations, agreements, laws, and specific guidelines have been adhered to.

component interaction matrices

Applied to improve understanding of indirect impacts from projects; the matrices identify first-, second-, and high-order impacts and illustrate the dependencies between environmental components.

comprehensive study EIA

An EIA assessment applied on large-scale, complex, environmentally sensitive projects that have a high risk of causing adverse environmental effects.

confirmatory analysis

Used to test for uncertainty in impact predictive techniques and to ensure similar predictive outcomes from different types of techniques.

consistency

A measure of the quality of an expert's judgement, including a test of the degree of randomness in a set of assessment scores or judgements.

continuous impacts

Impacts that are ongoing over space or time.

control site

A reference point where the environment is not affected by the project; used to monitor the nature and extent of project-induced environmental change in those areas that are affected by the project.

cost-benefit analysis

An assessment method that expresses project impacts based in monetary terms.

cumulative effects assessment (CEA)

The assessment of changes to the environment that are caused by actions in combination with other past, present, and future actions; changes may be linear, amplifying or exponential, discontinuous, or structural surprises.

cumulative effects monitoring

Non-site-specific monitoring that emphasizes the accumulated effects of multiple developments within a particular region.

decision-point audit
Examination of role and effectiveness of the EIS based on whether the project is allowed to proceed and under what conditions.

Delphi technique
An iterative survey-type questionnaire that solicits the advice of a group of experts, provides feedback to all participants on the statistical summaries of the responses, and provides an opportunity for each expert to revise her/his judgements.

descriptive checklist
Type of EIA checklist that gives guidance on how to assess impacts and what information is required.

deterministic model
A model dependent on fixed relationships between environmental components.

direct impacts
First order impacts that have a particular social value.

discontinuous effects
Effects that occur in an incremental manner and go unnoticed until some threshold is reached.

draft EIS audit
Review of the project EIS according to its terms of reference.

early warning indicators
Biological or non-biological indicators that can be measured to detect the possibility of adverse stress on VECs before they are adversely affected.

effects-based CEA
Measures multiple environmental responses to stressors; the results are then compared to some control point to determine the actual measure of cumulative change.

environment
Refers to all environments, including biophysical, social, and economic components.

Environmental Assessment Review Process
The first Canadian federal EIA process, formally introduced in 1973 by guidelines order.

environmental baseline
The present and likely future state of the environment without the proposed project or activity.

environmental effects
Measurable change in the state of the environment.

environmental impact assessment (EIA)
A systematic process designed to identify, predict, and propose management measures concerning the possible implications that a proposed project's actions may have on the environment.

environmental impacts
Environmental effects that have an estimated societal value placed on them.

environmental impact statement (EIS)
The formal documentation produced from the EIA process that provides a nontechnical summary of major findings, statement of assessment purpose and need, detailed description of the proposed action, impacts, alternatives, and mitigation measures.

environmental management system
A voluntary industry-based management system that controls the activities of products and processes that could cause adverse environmental impacts; management systems are a cyclical process of continual improvement where industries and firms are constantly reviewing and revising their way of doing business to protect the environment.

environmental preview report
An early report that is used to determine whether or not an EIA is needed and the level of assessment required; it outlines the project, potential environmental effects, alternatives, and management measures.

environmental protection plans
Mandatory management plans that result from project based EIAs; these management plans are tailored to the project due to the identification of key impacts and issues and management measures through the EIA process.

environmental systems
Environmental components functioning together as a unit.

exclusion list
A screening mechanism listing those projects subject to an EIA unless they are

included in the list; generally, projects excluded include issues of national defence and emergency, or those projects that are routine in nature.

experimental monitoring

Research into environmental systems and their impacts for the purpose of gathering information and knowledge and testing hypotheses.

ex-post evaluation

Taking action and making decisions based on the result of structure, analysis, and appraisal of information concerning project impacts.

Federal Environmental Assessment Review Office (FEARO)

The federal agency created to oversee the implementation of the Federal Environmental Assessment Review Process.

fixed-point scoring

A VEC weighting approach where a fixed number of values is distributed among all affected environmental components; the higher the point score the more important the environmental component.

fly-in fly-out

Projects located in remote areas with high amounts of air traffic to and from the site; typically associated with remote mining projects where temporary project work camps are constructed to house workers, who commute by charter air service on a basis, for example, of two weeks at the work site and two weeks off, during the lifespan of the project.

functional scale

Scale relationship based on how different environmental components function across space.

Gaussian dispersion model

A model devised for predicting point-source atmospheric pollution.

geographic information system (GIS)

A system of computer hardware and software for working with spatially integrated and geo-referenced data.

gradient-to-background monitoring

Monitoring system that measures the effects caused by the impact source, at an increasing distance from the impact origin to the point of background assimilation; an 'artificial' control point is established.

gravity model

A deterministic model used to predict population flow or spatial interaction; the model is dependent on a fixed and inverse relationship between mass (population) and distance.

holistic approach

Examining environmental components based on their position and functioning within the context of the broader environmental system.

human environment

Those aspects of the environment that are non-biophysical components; also referred to as the socio-economic or cultural environment.

impact avoidance

A form of impact management whereby impacts are avoided at the outset by way of alternative project designs, timing, or location, rather than managed or mitigated after they occur.

impact benefit agreement

Legal agreement between a proponent and a community or group that will potentially be affected by a project; generally applied to ensure that the resources for maximizing the benefits associated with the development are fully capitalized on.

impact indicators

Provide a measure of qualitative or quantitative magnitude for an environmental impact and might include, for example, specific parameters of air quality, water quality, or employment rates; allow decision-makers to gauge environmental change efficiently.

impact mitigation

Minimizing adverse environmental change associated with a project by implementing environmentally sound construction, operating, scheduling, and management principles and practices within project design.

impact significance

Reflects the degree of importance of an impact and is based on the characteristics of the impact, receiving environment, and societal values.

implementation audit
An evaluation of whether or not the recommendations presented in a project's EIS were actually put into practice.

inactive publics
Publics not normally involved in the environmental planning, decisions, or project issues.

inclusion list
A screening method listing those projects that have mandatory or discretionary requirements for an EIA.

incremental effects
Marginal changes in the condition of an environmental component caused by project actions.

initial environmental examination
Information prepared to establish whether or not an EIA is needed and what level of EIA should be implemented.

input evaluation
Follow-up or auditing of SEA procedures and requirements, such as purpose and objectives, data quality, and linkages to policy, planning, or program initiatives.

inspection monitoring
Site-specific monitoring with on-site visits to ensure compliance with procedures and safety standards.

intention survey
A survey that attempts to collect the judgement of as many people as possible and record their responses to what they intend to do or how they might react, given certain circumstances or situations.

interaction matrix
A type of EIA matrix based on the multiplicative properties of simple matrices to generate a quantitative impact of the proposed project on interacting environmental components.

ISO 14001
Internationally recognized industry performance standard for environmental management system certification.

Keynesian multiplier
A basic economic multiplier that notes that an injection of money into a local economy will increase at the local level by some multiple of that initial injection.

Leopold matrix
An EIA matrix for identifying first-order project-environment interactions, consisting of a grid of 100 possible project actions along a horizontal axis and 88 environmental considerations along a vertical axis.

life-cycle assessment
The 'cradle-to-grave' assessment of projects from their inception and start-up to post-operation.

linear additive effects
Incremental additions to or from a fixed environment, where each additional increment has the same effect.

list based screening
A checklist of projects that may or may not require an EIA.

Mackenzie Valley Environmental Impact Review Board
A Valley-wide public board created as part of the Mackenzie Valley Resource Management Act to undertake EIAs and panel reviews under the jurisdiction of the Act.

Mackenzie Valley Resource Management Act
An Act implemented by the federal government to give decision-making authority to northerners concerning environment and resource development activities within the Mackenzie Valley region of the Northwest Territories; proclaimed in 1998, the Act governs EIA in the region.

magnitude
The size or the degree of a predicted impact; not to be equated with significance.

magnitude matrices
Matrices that attempt to identify impacts and summarize impact importance, time frame, and magnitude.

matrices
Management and assessment tools used for identifying project impacts that typically consist of a two-dimensional checklist that places project actions on one axis and environmental components on the other.

maximum allowable effects level
An approach to impact prediction based on specifying certain desired limits or

thresholds beyond which a certain impact is not to exceed.

mechanistic model
A type of model based on mathematical equations or flow diagrams that describe cause-effect relationships in a project environment.

mediation EIA
An approach to EIA where an independent neutral mediator helps to resolve conflict between different interests groups, stakeholders, and proponents throughout the assessment.

methods
The various aspects of an assessment, including organization, identification of impacts, and the collection and classification of data.

model-class screenings
Provide a generic assessment of all projects within a certain class, where information contained in a 'model' report is prepared for individual projects and adapted for location or project-specific information.

models
Box-and-arrow or mathematical equations used to simplify real-world environmental systems.

monitoring
A systematic process of data collection or observations used to identify the cause and nature of environmental change.

monitoring of agreements
Monitoring and auditing of agreements between project proponents and affected groups to ensure compliance.

monitoring for knowledge
The monitoring used after impacts occur; data are collected and used for future impact prediction and project management.

monitoring for management
Tracking and evaluating changes in a range of environmental, economic, and social variables; usually associated with high-profile projects with uncertain outcomes and with the potential for significant adverse outcomes.

Monte Carlo analysis
A means of statistical evaluation of mathematical functions using random samples.

multiplier
A quantitative expression of some initial exogenous change and the expected additional effects caused by interdependences with an endogenous linkage system.

National Environmental Policy Act (NEPA)
The US legislation of 1969 that required certain development project proponents to illustrate that their projects would not cause adverse environmental effects; the beginning of formal EIA.

networks
An EIA method used to identify direct and indirect impacts that may be triggered by early project activities.

Nunavut Impact Review Board
Established under the Nunavut Land Claims Agreement and the primary authority responsible for EIA in the land claims area.

Nunavut Land Claims Agreement
Canada's largest land claims settlement and land claims-based EIA process; signed in 1993 giving the Inuit self-governing authority, and leading to the establishment of a new territory, Nunavut, in 1999.

off-site impacts
Impacts that occur at a distance or removed from the recognized project area.

on-site impacts
Impacts that occur directly in the immediate project area.

open scoping
A transparent scoping process where the content and scope of the assessment are determined from consultation with various interests groups and public stakeholders.

output evaluation
Monitoring or auditing SEA effectiveness based on the output of the SEA process, namely whether the process informed policy and whether the intended outcomes were actually realized.

paired comparisons
An approach to determine the relative importance of impacts in a hierarchy in ratio form, where the decision-maker considers trade-offs one at a time for each pair of VECs.

performance audit
An assessment of a proponent's capability to respond to environmental incidents and of its management performance.

Peterson matrix
A multiplicative EIA matrix consisting of project impacts and causal factors, resultant impacts on the human environment, and relative importance of those human components used to derive an overall project impact score.

phenomenon scale
The scale used to determine the spatial extent within which certain environmental components and VECs operate and function.

plan
A defined strategy or proposed design to carry out a particular course of action or several actions, and various options and means to implement those actions.

policy
A guiding intent, set of defined goals, objectives, and priorities either actual or proposed.

policy-based SEA
Strategic environmental assessment applied to policies that have no explicit 'on-the-ground' dimension, such as fiscal policies or national energy policies; also referred to as indirect SEA.

precision
The exactness of impact prediction.

predictive technique audit
A type of environmental audit where a project's predicted effects are compared to the actual effects.

probability analysis
An analysis that uses quantified probability to classify the likelihood of an impact occurring and under what environmental conditions.

process evaluation
Monitoring and auditing SEA based on the actual SEA process, including methods , techniques, openness, and frameworks.

program
A schedule of proposed commitments or activities to be implemented within or by a particular sector, plan, or area of policy.

programmatic environmental assessment
The application of project-based EIA to area-wide project or program management and development initiatives.

programmed-text checklist
A type of EIA checklist consisting of a series of filter questions for project screening and impact identification; useful for standard or routine projects.

project evaluation monitoring
Performance auditing or productivity measurement, a monitoring program concerned with a project's performance and ability to reach specified goals and objectives.

project impact audit
A type of auditing that focuses on determining whether the actual project impacts were predicted in the EIS.

questionnaire checklist
An EIA screening method consisting of a set of questions that must be answered when considering the potential effects of a project.

rating
An approach to VEC or impact weighting whereby the importance or significance of each is indicated on a numerical scale ranging from, for example, 1 to 5; no direct decision trade-offs are involved.

rectifying impacts
An approach to impact management based on restoring environmental quality, rehabilitating certain environmental features, or restoring environmental components to certain degrees following an environmental impact.

regional CEA
Cumulative effects assessment applied over a broader spatial scale to assesses a wide range of impacts in a specific region or area.

regional SEA
An approach to SEA concerned with regional-based environmental planning or development, and assessing the impacts of area-specific plans and program initiatives.

regulatory permit monitoring
Site-specific monitoring that includes regular documentation of requirements necessary for permit renewal or maintenance.

relative significance

A measurement of the relative importance of one affected environmental component over another.

replacement-class screening

A generic assessment of all projects within a class, but no location- or project-specific information is required and thus a screening report is not necessary for each individual project.

residual impacts

Impacts that remain after all management and mitigation measures have been implemented.

restrictive bounding

An approach to project scoping where the spatial area identified for consideration in the impact assessment is perhaps too small for a complete understanding of total environmental effects and within which the impact of the proposed project may be inflated due to the small area under consideration.

review panel EIA

A level of EIA that is applied to projects with uncertain or potentially significant effects or if public and stakeholder concern warrants an independent review panel.

risk

The possibility that an undesired outcome may result from an uncertain situation.

risk assessment

Using collected information to identify potential risks.

sanitary landfill

A type of waste disposal site where solid waste is contained within an impermeable barrier within the earth's surface and covered.

scenario analysis

An approach used in EIA and SEA to identify hypothetical actions or situations and potential outcomes.

scoping

An early component of the EIA process used to identify important issues and parameters that should be included in the assessment.

screening

The selection process used to determine which projects need to undergo an EIA and to what extent.

screening EIA

A type of project assessment that is an extension of the basic screening process where anticipated environmental effects are documented and the need for additional project modification or further assessment is determined.

secondary impacts

An impact resulting from a direct impact.

sector-based SEA

A type of SEA based on initiatives, plans, and programs that are specific to certain industrial sectors.

sensitivity analysis

Examining the sensitivity of an impact prediction to minor differences in input data, environmental parameters, and assumptions.

socio-economic monitoring

A type of monitoring that looks specifically at socio-economic parameters in the review of a project area.

Sorensen network

A hybrid of an EIA matrix and simple network that identifies direct impacts triggered by project actions and cause-effect relationships.

spatial scale

The actual geographic scale used to define the extent of a project EIA.

statistical model

A type of model used to test relationships between variables and to extrapolate data.

statistical significance

The determination, based on confidence intervals and probabilistic data, if a particular outcome or prediction is significant based on theoretical and empirical findings.

stochastic model

A type of mechanistic model that is probabilistic in nature or gives an indication of the probability of an event occurring within specified spatial and temporal scales.

strategic environmental assessment (SEA)

The environmental assessment of initiatives, policies, plans, and programs and their alternatives.

stressor-based CEA
Assessment that predicts cumulative effects associated with a particular agent of change.

structural surprises
Cumulative effects that occur in regions with multiple developments; these are the least understood and the most difficult cumulative effects to assess.

synergistic impacts
Impacts where the total effects are greater than the sum of the separate, individual effects.

systems diagrams
Models based on box-and-arrow diagrams that consist of environmental components linked by arrows indicative of the nature of energy flow or interaction between them.

techniques
Ways of providing and analyzing data in EIA.

threshold-based prediction
Basing impact predictions on prior experiences using approaches such as maximum allowable effects levels, where an impact is capped and not to exceed a certain threshold or level of change.

threshold-based screening
A screening process whereby proposed developments are placed in categories and thresholds set for each type of development, such as project size, level of emissions generated, or area affected.

threshold of concern checklist
A type of EIA checklist that lists environmental components that might be affected by the project actions, specific criteria for each component, and thresholds against which the project actions can be assessed.

traditional knowledge
Local or Aboriginal knowledge acquired from experience, culture, or interaction with land and resources over time.

valued ecosystem components (VECs)
Those components of the human and physical environment that are considered to be important and that therefore require evaluation within EIA.

VEC information
Information pertaining to the processes resulting from project-induced effects, such as habitat fragmentation, that is important to consider when characterizing the environmental setting.

VEC objectives
The specific parameters, guidelines, or standards set for potentially affected VECs.

weighted impact interaction matrices
An EIA matrix method where impacts are multiplied by the relative importance of the affected environmental components and secondary impacts are explicitly incorporated.

weighted magnitude matrices
An EIA matrix method where degrees of importance, representing the potential impacts of a particular project action on an environmental component, are assigned to the affected environmental components and then multiplied by project impacts.

Index